CHOKED

CHOKED

Life and Breath in the Age of Air Pollution

BETH GARDINER

The University of Chicago Press

Chicago

The University of Chicago Press, Chicago 60637
© 2019 by Beth Gardiner

Printed in the United States of America

28 27 26 25 24 23 22 21 20 19   1 2 3 4 5

ISBN-13: 978-0-226-49585-9 (cloth)
ISBN-13: 978-0-226-63079-3 (e-book)

DOI: https://doi.org/10.7208/chicago/9780226630793.001.0001

Library of Congress Cataloging-in-Publication Data

Names: Gardiner, Beth, author.
Title: Choked: life and breath in the age of air pollution / Beth Gardiner.
Description: Chicago: The University of Chicago Press, 2019. | Includes
    bibliographical references and index.
Identifiers: LCCN 2018052335 | ISBN 9780226495859 (cloth: alk. paper) |
    ISBN 9780226630793 (e-book)
Subjects: LCSH: Air—Pollution. | Air—Pollution—Social aspects. | Air—
    Pollution—Health aspects. | Environmental quality.
Classification: LCC TD883.1 .G37 2019 | DDC 615.9/02—dc23
LC record available at https://lccn.loc.gov/2018052335

♾ This paper meets the requirements of ANSI/NISO Z39.48-1992
(Permanence of Paper).

For Dan and Anna

CONTENTS

PROLOGUE

INHALE

The Meaning of a Breath

A human breath begins in the deepest reaches of the brain, where—far beneath consciousness—the body's most basic and essential functions are regulated. Just above the point where the spine meets the skull, tiny receptors detect rising levels of carbon dioxide, then stimulate nearby clumps of neurons. Between 12 and 20 times a minute, perhaps 20,000 times a day, millions of times a year, over and over and over again from the first cry of birth until the very last moment of life, those neurons fire signals ordering the muscles of the diaphragm and rib cage to contract.

Message received, the dome-shaped diaphragm flattens, and the ribs move upward and out. As the chest cavity expands, the pressure within drops, drawing air through the nose and mouth. Down the back of the throat, over the voice box, it follows its path deeper and deeper into the body.

To the naked eye, the lungs are unremarkable hunks of spongy pink tissue. Only when they inflate, puffing up like balloons, but faster and more dramatically, does their uniqueness become apparent. In the stylized illustrations of medical textbooks, a lung looks like an upside-down tree, a symmetrical maze whose branches get smaller and smaller as they divide into more than a million tiny twigs.

It's up close that the structure's elegant intricacy comes into focus. Hunched over his microscope, the seventeenth-century Italian anatomist Marcello Malpighi was the first to get a glimpse. Until then, the best medical minds believed air mixed directly with blood inside the lung. What Malpighi discovered is now a biological commonplace,

1

taught to middle schoolers: Inhaled air fills sacs that cluster at the airways' ends like miniature bunches of grapes. There are some 300 million of these alveoli in a pair of lungs, and their surface area is often likened, in total, to the size of a tennis court. Separated from them by membranes one one-hundredth as thick as a hair, tiny capillaries carry blood low in oxygen and laden with carbon dioxide. The gases rush across the barrier, and oxygen molecules bind to hemoglobin, then whoosh toward the heart, ready to be delivered wherever they are needed.

Like so much about the body, a breath is at once astonishingly simple and magnificently complex, delicately balanced yet highly resilient. Unlike other essential functions, the beating of the heart or the peristalsis of digestion, breath can also be controlled by the conscious mind, when we laugh or speak or hold it in to dive underwater.

The lung, too, is a point of vulnerability. While it has its defenses— the mucus that traps some contaminants, the hairlike cilia that sweep away others—this is the place where the outside world makes its way into the very center of the body, barriers left far behind as the air and whatever it carries come within a whisper of the bloodstream.

Lacking microscopes and an understanding of air's composition, the ancients struggled to grasp the hows and whys of a function they could plainly see was essential. Aristotle, a physician's son, believed breathing released heat generated by the vital fires of the soul. It would take millennia to definitively correct such misconceptions, but there was one fact the philosopher and his contemporaries understood well: Our very existence depends upon air. "The last act when life comes to a close is the letting out of the breath," wrote Aristotle. "And hence, its admission must have been the beginning."[1]

* * *

I didn't think much about the meaning or mechanics of breath when I was growing up. Until I was five, I lived in Fort Lee, New Jersey, in an apartment building overlooking the entrance to the George Washington Bridge. It was, and still is, a notorious traffic choke point, where the New Jersey Turnpike and a tangle of other highways merge into one. Suburbanites commuting to offices in New York sit bumper to bumper with trucks rolling north along the I-95 East Coast corridor, waiting to cross into the city. This was the early 1970s, the dawn of environmental consciousness, a time when American cars were exponentially dirtier than today, their fuel tainted with lead, sulfur, and other dangerous toxins. Along with my peers—the Jennifers, Davids, and Lisas who lived in the building too—I breathed it all in as we ran and jumped in the concrete playground out back.

Years later, in my twenties, I watched crosstown buses lurch by a few stories below my Manhattan apartment, heading into and out of Central Park. When I cleaned, the soot that coated the windowsill blackened big wads of paper towels. It took several goings-over, and a lot of soapy spray, to get the paint on the sill white again.

At 30, I moved to London with the charming Brit who'd stolen my heart. For 18 years, more than half of them with our chatty, energetic daughter, we've breathed the diesel fumes that foul the city's air. London's pollution is just one piece of a health disaster playing out across Europe, belying the continent's reputation for environmental progressivism. I can smell the exhaust when I'm out running errands, meeting a friend for coffee, or walking Anna to school, clouds of it billowing into our faces. After a few minutes on a busy road, I often have a mild headache.

For a long time, I saw that awful air as just an annoyance. But as I've come to understand pollution's profound effects on the human body, it's grown into something more: the focus of nagging worry, a fear for my daughter's well-being, and my own.

I try my best to protect her, of course. But I can't change the air that surrounds us, nor wish away our city's diesel mess. So I find myself veering from anxiety to anger. Landing sometimes, too, at a willful blocking out of the danger, as a parent's urge to shield a vulnerable child tugs against the reality of an individual's powerlessness in the face of larger forces. It's that back-and-forth pull of emotions that set me on the path to writing this book.

What I've found since is that the taint my family and I breathe is only one strand of a far bigger story. Around the world, from Fort Lee to Frankfurt, Karachi to California, dirty air causes 7 million early deaths annually,[2] more than AIDS, diabetes, and traffic accidents combined, making it the single biggest environmental threat to health.[3] New data suggests that number may climb even higher, pushing air pollution into the very top tier of global killers.[4] More than 40 percent of Americans breathe unhealthy levels of pollution.[5] In Britain, air pollution is second only to smoking as a health risk, causing as many as a fifth of all deaths in my adopted hometown.[6] Across Europe, it kills more than 15 times as many people as car crashes.[7] Nothing is as elemental, as essential to human life, as the air we breathe. Yet around the world, in rich countries and poor ones, it is quietly poisoning us.

It's not just the obvious ailments like asthma and bronchitis. Over the past decade and a half, scientists' understanding of air pollution's harms has advanced rapidly, and a powerful body of evidence now links it to a long and growing list of health woes, including heart attacks, strokes, birth defects, many kinds of cancer, dementia, diabetes, and Parkinson's disease.

Not far from where I grew up, and many years later, researchers took advantage of the natural experiment created when highways in New Jersey and Pennsylvania replaced the old-fashioned tollbooths where we used to hand cash to an attendant or toss coins into a basket with the high-tech kind that charge the fee electronically as you zoom past. The switchover dramatically reduced backups at the collection points, and the researchers found rates of premature birth dropped by about 9 percent for pregnant women within a mile and a quarter radius of where the old booths had been.[8] While that change was positive, it made clear the disturbing link between the highway exhaust so many of us breathe and a pregnancy outcome that can have lifelong consequences for babies born early.

I'm lucky enough to have been in good health all my life. And when, sooner or later, illness comes along, I won't know whether the pollution I've taken in over the years has had anything to do with it, or if I should, instead, blame my sweet tooth, my lifelong preference for a good book over a brisk run, or a malevolent gene buried somewhere in my DNA, beyond anyone's control.

None of us can. And that invisibility is a strange feature of this crisis. "You see one person run over in the street and you'll never forget it," observed a Los Angeles environmentalist I met. But thousands dying from the effects of dirty air "will never even faze you."[9] He was right. When smokers succumb, they know their own actions, and those of the tobacco companies that fed their habit, helped bring about their illness. But, in a world powered by fossil fuels, we all travel from place to place, use electricity, heat our homes, and few of us fully grasp the effects.

The gains that come when air gets cleaner are similarly difficult to see. There's no doubt reducing pollution saves lives. But those whose years are lengthened, and those who love them, never know it. Emergency room visits are averted and health care dollars stay in pockets, but the line connecting car or power plant regulations with the size of medical budgets isn't easy to make out.

Once I'd grasped the dimensions of this hiding-in-plain-sight threat, I wanted to see, up close, how it was playing out around the world. And why. Is air pollution an inevitable part of modern life, something we must resign ourselves to living with? Or are there more malign forces at work, too, keeping us wedded to the old, dirty ways of doing things when better alternatives exist? And, most importantly, what would it look like to do things differently, to build a cleaner, healthier world? Has anyone done it, or tried? And how can we get from here to there?

By way of looking for answers, this book tells the story—the stories—of air pollution, and of the people I met whose lives are shaped by it. In the United States, thanks to decades of gradually tightening regulation, air is far cleaner than it once was. But that improvement—now at risk as the rules that brought it about come under assault—has failed to keep up with the science, which tells us more clearly with each new study that even relatively low pollution levels do real damage. Around the world, in fast-growing South Asia and China, in Europe's coal-burning east and its diesel-dependent west, in Cairo and Johannesburg and Lagos, the problem is much worse.

In Part 1, "Holding Our Breath," our itinerary includes Delhi, pollution's ground zero, where a mother has nightmares about what's happening inside her children's lungs, and an eccentric businessman builds himself a clean-air bubble. I'll show you around London, my window onto the consequences of Europe's disastrous embrace of diesel. Poland's story is the story of coal; we'll meet men hoisting bags of it into their cars and watch an old woman trudge to the basement to scoop some into her furnace. In California's parched and poor San Joaquin Valley, migraines and wheezing shatter preconceptions that this is just an urban problem. We see there, too, that while dirty air affects everyone who breathes it, some suffer more than others, so this issue is infused with questions of race, class, and fairness.

The news is not all bad. Part 2, "Coming Up for Air," finds progress and hope—past, present, and future; grinding and swift; tiny glimmers and potentially seismic shifts. From the still-unfolding story of one of America's most powerful and important laws, the Clean Air Act—and the friendship that, across party lines, helped to birth it—to China's world-changing push to find a healthier path. And a search, on the streets of Berlin, for ways to put the needs of human beings before those of the cars that dominate so many of our cities.

Of course, there is action and progress in the polluted places, and still plenty fouling the air of the improving ones. My hope is that these chapters' global reach will demonstrate both the scale of the problem and the very real opportunity we have to solve it.

In many ways, this is a book about choices. About how we choose the kind of world we want to live in. And about the complexities of an age in which the things that have changed our lives for the better also bring consequences that are harder to see.

Cleaner air, it turns out, is not an impossible dream. We know how to get there, and doing so would bring enormous health benefits, on a par in some places with slashing sugar consumption or getting everybody up off the couch.

And understanding pollution's hidden dangers holds an even greater power. The overarching challenge of our time, of course, is climate change. But despite the floods and droughts and storms, its risks can still feel abstract and distant, a calculus of parts per million high in the atmosphere, or glaciers melting thousands of miles away.

Dirty air, on the other hand—caused by our heedless burning of the very same oil, gas, and coal that are warming the planet—is wreaking its damage in the here and now. Those fuels are woven into the fabric of our societies, and weaning ourselves from them won't be quick or easy. But once we recognize the toll they are taking, not just on the habitats of polar bears but on our hearts and lungs and those of our children and parents, I hope we'll see more clearly that a different future is within our grasp.

PART ONE

HOLDING OUR BREATH

1

THE MEASURE OF A LUNG

Charting Pollution's Power

In a hospital, a bedroom, on a mat laid out on a dirt floor, with a groan and one last push, a child is born. The light, the cold, the noise come as a confusing shock after the warmth of the womb, but one thing is clear and pressing, unfamiliar but primal: a need for air. The fluid that filled the lungs for months has begun to seep away. Now their work must begin—the work of a lifetime, the work of sustaining life. No longer can this child rely on her mother's lungs, her mother's heart, her mother's blood to deliver the oxygen she needs, to whisk away the carbon dioxide that accumulates so quickly. She depends on others in almost every way, but this she must do for herself. Her first try, the first breath, is a gasp, shallow and difficult. But it makes the next one a bit easier, squeezes out a little more of the fluid, inflates the tiny alveoli where oxygen will cross to the bloodstream. Unsteady at first, the new rhythm soon takes hold, the newborn's chest rising and falling, each breath expanding the lungs a little more. Even deeper within the child's body, the rush of oxygen resets the pressure in her blood vessels. That change cascades toward her heart, shifts the balance of its pressure, too, soon forces closed a hole that had connected left to right. The old map is redrawn, and the blood follows a new path, as the lungs—until this moment irrelevant, now utterly essential—become a stop on its journey through the body. The breath is getting easier, in and out, again and again, as life in this remarkable new world begins.

Across Southern California, in school gyms and libraries and lunch-rooms, the children filed in, one by one, to put their lips around a plastic tube and blow with all their might. Thousands of them, year after year, in rich neighborhoods and poor ones, from the breezy towns along the Pacific coast to the hot, smoggy valley locals know as the Inland Empire. Erika Fields was one of them, back in the 1990s, when she was in high school at Long Beach Poly, just outside Los Angeles. Even now, she's the kind of person who raises her hand, who steps forward when volunteers are needed, and she liked being the only one called out of her class, walking down the hall to the quiet room where the breathing machine sat on a desk. She liked, too, the sense of being part of something bigger than herself, something that might really matter in the world.

In the empty classroom, the woman from the University of Southern California would hand her a sterile mouthpiece, attached by a tube to the spirometer ready to gauge the power of her lungs. Erika would give it a couple of practice puffs to get comfortable before the one that counted. "I remember her saying 'Push, push, push. Blow all the air out.' And then she would show me on her laptop, and I could see on a graph where I pushed the most," and watch the line edge downward as her breath tailed off. After that, there was a survey to fill out, a couple of pages about her health and her family, about smoking in the home and pets and diet and exercise, and then Erika would walk back down the hall, back to her classmates and the ordinary rhythms of the school day.

She didn't know it then, but those brief once-a-year interruptions to her routine helped lay the foundation for insights that would ultimately change scientists' understanding of what air pollution does to the human body. In the vast stacks of accumulating numbers—results from Erika Fields's breath tests and thousands of others—a team of patient researchers would discern the outlines of a threat that had, until then, been hard to see.

* * *

Ed Avol was one of those scientists. He grew up breathing the foul air of 1960s L.A., and he remembers well the hacking coughs that filled the playgrounds of his childhood. An engineer by training, he worked early in his career on hospital-based studies that examined the effects of dirty air as researchers had for decades, by pumping pollution into small rooms and watching volunteers exercise inside. The team he was part of wasn't allowed to make conditions in their smog chambers any worse than what Angelenos would experience outdoors, but in the 1980s that still gave them plenty of latitude. The researchers would monitor subjects as they pedaled, measuring their heart rates and oxygen levels, making note of their coughing, their shortness of breath, and their red, watery eyes. By that time, it was clear to scientists that ozone—the main ingredient in the smog that still plagues L.A. and so many other cities—had an immediate effect on those who breathed it. And the impact could be far more serious than the discomfort Avol saw so plainly: When ozone blankets a city, asthmatics wheeze, emergency room visits spike, and even in healthy people, the lungs can grow inflamed and struggle to do their job.

But Avol had begun to ponder an even bigger question: If ozone's immediate effects on the body were so clear, what was it doing over the long term to those who breathed it day after day, month after month, from one year to the next, over the course of a lifetime? Did its effects vanish when the air cleared, or had ozone—or, perhaps, some other, less familiar pollutant—wrought unseen damage that would accumulate slowly, lying in wait to bite unexpectedly decades later?

As it happened, John Peters and Duncan Thomas, researchers at USC's medical school, were wondering the same thing. Concerned with the implications for public health, California's Air Resources Board, the country's most aggressive air pollution regulator, had asked them to design a long-term study that would definitively answer that very question. Avol signed on, and the team began, in the early 1990s, to set out the parameters that would guide their effort, to ensure that its long, slow time frame would be rewarded by results of rock-solid reliability.

It was that design that, a few years later, drew Jim Gauderman to the project, known by then as the Children's Health Study. Gauderman is a biostatistician, schooled in the complex methodologies that sift human truths from vast haystacks of numbers. It's not just his training that suits him to his work. He speaks in measured tones, weighs his words thoughtfully, has the calm, steady bearing of a man who knows painstaking diligence is sometimes the only route to scientific clarity.

He could see the meticulous groundwork Avol and the others had laid would one day bring a clear answer to a question that ate at him too. Gauderman was also from L.A., and had wondered, since his days as a young cross-country runner, what hidden scars all those training sessions along traffic-clogged roads might have left on him. He and Avol, junior partners at first on a project others had conceived, would eventually become its leaders.

There were many metrics the scientists could have chosen to use, but the one they settled on carried a unique power—the power, in some sense, of life itself. They'd test the strength of children's lungs, over and over again, year after year. It was a measurement that, at least implicitly, carried a prediction, one that would follow a child forever. For there are few facts about a human body that correlate more closely with its health, with the very length of its life, than the development of its lungs. Weaker lungs mean illness and frailty, and fewer years; stronger ones allow for vitality and longevity. So what Avol and Gauderman were measuring, very literally, were the invisible boundaries being drawn around their subjects' futures.

For the children, of course, those stories were yet to be written. The team recruited 4,000 of them, kids between 10 and 16, in a dozen towns and neighborhoods as diverse as California itself. Later, they would add more cohorts, bringing the total number of subjects to more than 11,000, the youngest of whom joined as kindergartners and finally aged out of the study when they finished high school in 2015.

It was Gauderman who sifted, day after day, through the data that began pouring in. In addition to the annual lung tests, there were height and weight checks for every participant, and air quality measurements from monitors in the children's neighborhoods. To a man who knew how to read them, the numbers sketched the outlines of the children's lives and health, of their bodies and their lungs and the air that filled them.

It was still too soon, though, to find answers to the questions driving this work. While the years went by and the children's lungs grew, Gauderman would have to watch and wait. Early on, he was mostly looking for errors that might skew the results. Did the paperwork list a child's height as four feet when he stood last year at five? "We would go to the field staff and say, 'What happened? This kid shrunk a foot, which value is right?'"

Eventually, he began searching for patterns. They weren't easy to make out. Because, of course, all those young lungs were growing, and gaining power, no matter how dirty the air they breathed. Gauderman's task was to determine whether some were developing more slowly than others. If they were, there could be any number of causes, from genetic predisposition to poor diet or a parent who smoked. Even in the most polluted towns, there would be many children with strong lungs, and in the cleanest places, some would struggle for breath. He would have to use his most sophisticated statistical techniques, and sometimes invent new ones, to decode the message buried in the numbers, to disentangle what mattered from the irrelevant noise. As one millennium ended and another began, as new children joined the study and the older ones trudged down school corridors again and again, as their lungs matured and the data piled higher, the signal grew stronger. Until, eventually, it became undeniable.

There was no eureka moment, no sudden, made-for-TV "aha." It wasn't that kind of science, and Gauderman was not that kind of scientist. The message peering out at him wasn't what he had expected, so he checked and rechecked, looked at it from every angle, until he was certain what he was seeing was real. In 2004, the answer to the question the team had posed more than a decade earlier appeared, at last, on the pages of the publication reserved for the most consequential of health breakthroughs: the *New England Journal of Medicine*.

No one would have been surprised, of course, to discover pollution's effects were long lasting; they wouldn't have been looking if they hadn't had an inkling that might be the case. What shocked these careful men was the scale, the sheer force, of the impact. It turned out that, year by year, as their trove of data had grown, the damage was accumulating, too, creating a gap likely to forever haunt the children caught on the wrong side.

By the time the oldest subjects finished high school—when growth was winding down and the consequences had become irreversible— dirty air had left its mark on the lungs of the children who breathed it. Those lungs failed to reach their full power, were weaker than they ought to have been, the very air that had sustained them also holding them back. Pollution had planted a hidden seed of vulnerability in these children, an unseen frailty that would set their futures on a different trajectory than their peers', would make their bodies less robust and could even, eventually, shorten their lives.

Lungs that are 20 percent weaker than normal are a medical red flag, a warning that will prompt a doctor to hunt for a cause, and hope, especially when the patient is young, to diagnose something treatable. What the Children's Health Study found was that kids breathing the dirtiest air were nearly five times as likely to experience that level of lost function—almost 8 percent of them in the most polluted places, compared with 1.6 percent in the cleanest ones.[1] For every 100 children who grew up with the worst pollution, in other words, at least 6 would be burdened as a result with a lifelong health problem. Many more, of course, would have a lesser degree of impairment, a smaller wound that would, meaningfully if less dramatically, drain their vigor too. "That," Ed Avol recalls, "was a wake-up call."

The finding, Avol says now, shattered scientists' naïve assumption that if air pollution was doing serious, long-term harm, they would have known about it already. In fact, he explained to me, while the impact was profound, its slow, steady accumulation made it hard to spot. "That percent, percent and a half per year" of lost development, "unless you make careful measurements and have careful monitoring information and do some fairly sophisticated data analysis, it doesn't just jump out at you." Once they did so, pollution's power was plain to see.

It was the first glimpse of a frightening truth, one whose repercussions went far beyond Southern California. For the unsettling reality the team had uncovered was that, across the country and around the world, the fuels we have built our lives on—the gasoline, the diesel, the coal we burn to travel from place to place, to power our lightbulbs and laptops, to stay warm in winter—are changing our children's bodies.

There was another surprise in the data too. From the outset, ozone had looked like the prime suspect. Its immediate impact was so clear that if air pollution was doing any long-term damage, the scientists figured, the familiar noxious gas would be to blame. But while its harm was real, ozone was not what was dampening the power of those young lungs. Tiny airborne particles known as PM2.5, so small they are thought to enter the bloodstream and penetrate vital organs, including the brain, were a far more potent danger. Nitrogen dioxide, one of a family of gases known as NOx, also had a powerful effect. In fact, it poured out of cars, trucks, and ships in such close synchronicity with PM2.5 that even Jim Gauderman's statistical models couldn't disentangle the two pollutants' effects.

That wasn't all. In what may have been their most worrisome discovery, the team found the pollutants were wreaking harm even at levels long assumed to be safe. In the years to come, the implications of that uncomfortable finding would be felt far beyond the pages of prestigious scientific journals.

* * *

Long Beach, where Erika Fields grew up, and where she gave birth to a baby after high school, wasn't the most polluted of the Children's Health Study communities, but it was certainly on the wrong end of the scale. Around the time her son was born, she started getting bad colds that would last for days, then clear up in a matter of hours. Eventually, she saw a doctor, who diagnosed allergies and sinus problems and said poor air quality was probably to blame.

Today, Fields lives with her family in the suburbs east of Los Angeles, an area known for its terrible air, where the city's smog sits against the Santa Ana Mountains. The baby she had as a teenager is now a young man, and his sinuses act up just like his mom's. Her youngest son, an energetic grade schooler who climbs on the back of her chair as we speak on Skype, has asthma. The attacks don't happen often, but they're frightening when they do. It started suddenly, when he was a toddler. "He just kept vomiting and vomiting and got very lethargic, and panting like a puppy, and I could not understand what was going on," she tells me. It's never been quite that bad again, but "it's like he can't get a full breath of air in, so in trying to breathe in he's coughing." Even now that he's older, the struggle still causes him to throw up.

All these years later, the emails from USC are few and far between, but Fields replies to every one. Until we spoke, she hadn't heard about the study's findings, the breakthroughs in which she'd played a small role. She's still glad she raised her hand and walked down that hallway, though. "I felt," she tells me, "like I was just doing my part." With the help of Erika Fields and many others like her, anonymous but essential, the Children's Health Study had upended the modern understanding of air pollution, and set researchers around the world on a new path, one whose further reaches they are still charting today.

* * *

In the years since the study began publishing its results, the evidence on pollution has continued to accumulate, and the findings only grow more worrisome. Today, astonishingly, researchers tell us just 5 percent of humanity breathes truly healthful air,[2] and one out of every nine deaths on earth is caused by air pollution.[3] In America, air quality has improved dramatically since the days when I was breathing those thick fumes beside the George Washington Bridge. But as Ed Avol and Jim Gauderman found, even low levels cause real harm. So every year, dirty air still brings more than 100,000 American lives to a premature end.[4]

And those deaths come a decade earlier than they otherwise might have,[5] a terrible roster of missed birthdays and weddings, of grandchildren without grandparents and children without parents—sometimes, even, of parents without children. For each loss, we who survive may blame any number of causes—a loved one's flabby belly, his distaste for exercise, her fondness for wine or cigarettes. Or, sometimes, just plain bad luck. Almost never do we blame the air around us, but the truth is that it fells many.

By 2013, the weight of the evidence would grow strong enough for the World Health Organization (WHO) to deem air pollution a carcinogen. The following year, it more than doubled its previous estimate of the global death toll.[6] Now, the agency says outdoor air pollution kills 4.2 million people a year, while household air pollution, the scourge of developing nations where many cook over open fires, cuts short nearly 4 million lives (the tolls overlap, for a total of 7 million deaths).[7] Dirty air is the fourth largest risk factor for early death, far ahead of alcohol and physical inactivity.[8] And a major new analysis suggests the true toll may be even higher: it doubles the WHO's death estimate from outdoor sources to as many as 8.9 million,[9] meaning air pollution may be a bigger killer even than smoking. The toll can be measured in dollars as well as lives. Dirty air, says the World Bank, costs the global economy more than $5 trillion every year—the size of the economies of India, Canada, and Mexico combined.[10]

What the science shows is that the air we breathe affects every part of our bodies, not just our lungs, although of course it exacerbates asthma, causes lung cancer, and hastens the decline of those with other breathing problems. In fact, dirty air kills far more people through its cardiovascular effects—nearly three-quarters of the total—than its respiratory ones.[11] Even a few hours of exposure can have terrible consequences. High-pollution days bring more heart attacks[12] and strokes,[13] more emergency room visits—and more deaths. Children, the elderly, and people with existing illnesses are in greatest danger.

It's not just a temporary jump in risk. Pollution takes a cumulative toll on hearts and arteries, and the numbers are jarring. In one vast study, researchers tracked tens of thousands of healthy American women in their fifties, sixties, and seventies and found those breathing even moderate levels of particle pollution were 24 percent more likely to suffer a "cardiovascular event," the deceptively dry term for life-changing traumas like a stroke, heart attack, or bypass operation. Even worse, they were 76 percent likelier than women in less polluted places to die from such an event. And the dirtier the air they breathed, the higher the risk climbed.[14]

Scientists have gone deeper, too, searching for the hidden physical changes that presage such traumas, the biological mechanisms that move us down the path toward crisis. In one study, researchers used ultrasound scans to measure the arteries of subjects breathing dirty air, and saw them constricting after just 10 minutes.[15] Another team found blood clots increased by about a quarter among people exposed to diesel exhaust—and the changes lasted for at least six hours.[16] Calcification of the arteries, a powerful warning sign of impending trouble, rises along with pollution concentrations.[17] Even tests on people in their twenties showed pollution episodes coincided with damage to the cells that line the blood vessels.[18]

Researchers are still trying to unravel what drives such changes, exactly how dirty air wreaks its havoc. Experts now believe low-level inflammation is at the root of many of the problems that plague the human body, and it may be the key to pollution's power. There are other possible avenues too. Tiny particles may trigger changes in our immune systems, turning our bodies' protective armor against us, or reset genes and receptors, throwing their regulation of delicately balanced systems—the release of insulin, the communication among neurons—off-kilter.

The more I've learned about air pollution, though, the clearer it's become to me why, even in diesel-fouled London, no one I know talks much about the air. It is so hard to put the risks that surround us in perspective, to make them the right size. I know my chances of going down in a plane crash are infinitesimal, but that doesn't stop me from needing a glass of wine whenever a flight gets bumpy. Any one of us is similarly unlikely to be killed in a bombing, but it's a truism that terrorists spread fear far beyond their actual reach. We're hard-wired to respond viscerally to such threats, sudden and all too imaginable. That instinct served our cave-dwelling ancestors well, but in modern life, where risk is often more complex and difficult to apprehend, it sometimes leads us astray.

Unlike a pouncing tiger, the danger posed by pollution is slow moving and far from obvious. Even if we do take heed, it's hard to know what to do. That lack of control makes the threat especially upsetting. So, with no one around me acting as if each breath is drawing a little more poison into their body, I've done, time and again, what we do so often in the face of such threats. I try, when I can, to push it out of my mind and retreat back into ignorance.

But the science keeps moving on, and the list of maladies pegged to dirty air continues to grow. What I find most unsettling is the damage it does to the brain. Reams of evidence point to pollution as a trigger for cognitive decline, dementia, even mental illnesses like depression. Particles are believed able to seep through the blood-brain barrier and enter the brain directly from the nose, via the passageway of the olfactory nerve. In Taiwan, researchers found Alzheimer's rates jumping sharply as ozone and particle pollution increased.[19]

A particularly vivid bit of evidence came from a team of scientists who examined the brains of puppies from Mexico City, known at the time for its awful air. The markers they found—degenerating neurons, twisted protein fibers, plaque deposits—were the same telltale signs by which doctors diagnose Alzheimer's in humans.[20] That pointed not just to a link between dirty air and a dreaded ailment, but to the troubling possibility that the path toward dementia might be laid out in youth, not old age. That likelihood grew stronger still when the neuropathologist who led the dog study conducted autopsies on children and young adults killed in accidents and found early markers of Alzheimer's in 40 percent of those who'd lived with high levels of pollution, but none who had breathed clean air.[21] She saw, too, in the brains of young people, the red flags of Parkinson's, along with inflammation, impaired blood flow, and genetic changes,[22] all frightening harbingers of neurological decline. A follow-up study found immediate effects on children's function too. Kids who lived in the most polluted places and also carried a gene linked to Alzheimer's had short-term memory loss and IQs 10 points lower than their peers.[23]

It's not just the filthiest air that triggers such damage. A team studying American women in their seventies and eighties used MRIs to examine subjects' brains, and found that in those who lived with even slightly elevated pollution concentrations, the area known as white matter had shrunk as much as it would from a year or two of additional aging.[24] The effects are discernible even without expensive equipment. Another research group found that elderly women breathing pollution levels that are common in the United States did as badly on cognitive and memory tests as they would have if they were two years older.[25] Scientists find such changes in older people because that's where they've looked. I can't help wondering, worrying, whether they happen even earlier. Might my own brain—my husband's, my daughter's, our friends'—be shrinking, too, from our years of inhaling London's diesel stew?

What is perhaps most worrying is that the more scientists learn about dirty air, the clearer it becomes that there is no safe level. Even near the bottom of the scale—even far below the limits set by governments[26]—more pollution means more people die. In a landmark study of six American cities in the 1990s, a Harvard team was stunned by their discovery that pollution took two to three years off the lives of those in the dirtiest one—Steubenville, Ohio—even though its air met federal guidelines.[27] Another way of looking at that, the more hopeful flip side, is that putting less nitrogen dioxide, less ozone, fewer particles into the air, even in places that are relatively clean, saves lives almost immediately. And indeed, the Harvard researchers found in 2009 that death rates had fallen along with pollution levels in all their study cities, even the cleaner ones.[28]

*　*　*

In chunky black glasses and a patterned scarf, her dark hair pulled back, Beate Ritz still looks more the sophisticated European than the casual Californian, even after decades in America. Sunshine streams through a window into her home in the Santa Monica Mountains, above Los Angeles, as we speak on Skype, and she pours herself a cup of tea. Ritz is an epidemiologist at UCLA, and she knows it can be nearly impossible to link one individual's health problem to a specific environmental cause. But the work that would shape her career began with a nagging, personal worry.

The smog blanketing L.A. came as a foul shock when she arrived from her native Germany. She was expecting her first child, and the pregnancy was smooth and easy; she swam regularly right until the end. So when her son was born surprisingly small, the only explanation she could come up with was that exhaust from busy I-10, which passed right above her apartment, must have been to blame. The fear piqued her professional interest, and she began scouring research journals for hints on what the air a woman breathed while pregnant might do to her baby. It was 1990, and Ritz was dismayed to discover there had been almost no studies on the question. Answering it would become her life's work. But first, she had to protect her family, so she and her husband found a new home far from L.A.'s smoggy center. By the time her second son was born, they'd moved to the coastal enclave of Topanga. He was two pounds heavier than his older brother had been.

Second babies are often bigger than their older siblings, and Ritz knew she couldn't draw any wider conclusions. "That's anecdotal, and as a scientist you wouldn't believe it at all." So, as her little boys grew into young men, she devoted herself to the hard, slow task of gathering the data that would invest her hunch with the authority of science. For she knew that it is not just at the end of life, but also its beginning— before the beginning, even—that the human body is most vulnerable. She would search for the roots of disease in those early months when we are shielded from the world, our lungs not yet drawing breath.

She met resistance along the way. Some in her field doubted there could be any link between air quality and the well-being of a fetus, let alone the health of the child it would become. It was easy to believe pollution could affect a woman's lungs, the reviewers considering one of Ritz's early grant applications said, but they couldn't credit the notion that it might travel deeper within her body, deep enough to reach the baby she carried. An early hint those judges were wrong came when Ritz learned a colleague in Prague had found the molecular fingerprint of central Europe's ubiquitous coal smoke in the placentas and umbilical cord blood of newborns. If contaminants from Czech coal could reach growing fetuses, she knew, American car exhaust could too. The next question was, what did it do to them?

The answers, she believed, could be found in the birth certificates of tens of thousands of Californian babies and in the state's registries of congenital defects and childhood cancers. Ritz broke out each child's vital statistics by address, then matched the addresses against local air pollution levels. The analyses took weeks to run on the computers of the 1990s, but got quicker as the technology gained power.

Eventually, she found what she was looking for. She tells me how, gradually, she discovered one ailment after another was more prevalent in the children of mothers who had breathed dirty air while pregnant. The particular worry that had started her on this path was borne out: Underweight babies were 10 percent more common for women living near heavy traffic. So were premature births; worse still, extremely premature births were 80 percent more likely. Those findings have serious ramifications, because prematurity and low birth weight are both linked to health problems later in life. Ritz found risks of pre-eclampsia, a potentially serious pregnancy complication, increased with pollution levels too. When I ask whether she now blames pollution for her own son's small size, her reply reflects the scientific rigor that has always guided her: "It's unknowable."

Other researchers have linked miscarriage and infertility to pollution.[29] That latter fact touches a painful scar for me, a reminder of the years we spent, and the endless medical appointments, trying to conceive the second child I'd always imagined having, the brother or sister Anna never got. It's a sadness I've mostly put behind me now, grateful for the healthy, happy kid whose mom I do get to be. But there was a time, not so long ago, when stumbling across a study linking infertility to the kind of pollution my husband and I have breathed for years might have sent me back down a fruitless trail of anger and grief. London's air, of course, may have had nothing to do with our troubles, with the lost dream of that longed-for baby. Some things, as Beate Ritz says, are simply unknowable.

That unknowability, the statistical language in which pollution's dangers are inevitably framed, unmoored from any one life, strikes me as yet another reason they are so hard to grasp. Because we don't care, really, about percentages. We care about people, the people we live with, the people we love. So air pollution has a statistically significant relationship with infertility, with cardiovascular disease. What do I do with that? Is it why I have one child, not two? Did it cause the heart attack that terrified us, years ago, but from which my father, thank-fully, recovered fully? There are no answers to those questions. Most of us are not epidemiologists, trained to accept such uncertainty. But so many of the risks in our complicated world present just like this: real, and well documented, shouting for our attention from the headline of a story someone posts online. Yet intangible, diffuse, impossible to pin down. It's a strange, and very modern, kind of anxiety.

That is not to say, of course, that the statistics don't matter. When it comes to air pollution, they matter immensely, even when the danger appears, at first blush, to be small. Because the unforgiving logic of mathematics means that when even an innocuous-sounding, single-digit bump in risk is spread over an entire city, region, or nation, more people will draw an unlucky hand than when a health gamble whose odds look scarier is taken by a smaller group, like those who smoke or drink heavily. Ten percent may not sound so bad, but if an extra 10 percent of all babies in the polluted parts of L.A.—or Louisville, or Lagos—are born early, as Beate Ritz found, that adds up to a lot of premature babies. Because we all breathe tainted air, billions of us, just about everywhere, even a tiny upward tick in the incidence of a given ailment translates into vast numbers of victims. The only mystery, really, is who they are.

When the time came to present her first findings, Ritz worried how her colleagues might respond. "I was scared to death to have reviewers tell me again that the fetus does not breathe, and what was I doing," she recalls. As it turned out, she got the opposite response. Scientists were waking up to pollution's dangers, and they were eager to hear what she had found. And while Ritz did her work, others had begun similar studies. "Within a few years, it really mushroomed and you could find papers from all over the world reproducing what I've seen in L.A." Many of those studies have been done in China, a nation now engaged in the painful task of grappling with dirty air's effects on its children.

As Ritz pressed on, her findings grew more troubling. Low birth weight and early arrival are relatively common. Cancer in children, thankfully, is rarer, so she had to draw on nearly two decades of records to draw firm conclusions. She concentrated on those diagnosed at five or younger, the cases in which, if a connection to some prenatal exposure existed, it would be easiest to trace. What she discovered was chilling. Pediatric leukemia, kidney cancer, eye tumors, and malignancies in the ovaries and testes of young girls and boys were all more common in children whose mothers breathed traffic exhaust during pregnancy.[30] Diving deeper into the records, she found, too, that babies' death rates—from breathing problems and the unexplained tragedy of SIDS, sudden infant death syndrome—climbed along with pollution levels.[31] Ritz had tied the exhaust floating from millions of tailpipes to the worst kind of grief.

She found, too, that heart malformations—tiny, imperfect valves and aortas, holes in the cardiac wall—were three times more common in the children of mothers who had lived with pollution early in pregnancy, when the fetal heart is taking shape.[32] More recently, Ritz added a new wrinkle to the struggle to understand the causes of autism with her finding that women who breathe polluted air while pregnant are more likely to have an autistic child.[33]

Scientists examining dirty air's effects on adults can look to the bodies of mice and other lab animals to unpick the mechanisms of harm. But those creatures' pregnancies are too dissimilar to ours to offer useful insights, so we know little about exactly how pollution's damage is wrought in utero. Its invisible work may be done as early as the days after conception, when two sets of DNA twist together into one. Toxins might penetrate tiny, developing organs a bit later. Or they could interfere with development indirectly, by sparking reactions in the mother's immune system that have a domino effect on the growing baby.

It's entirely possible that the changes that pollution inscribes before birth haunt us not just during childhood, but throughout life. It would take decades-long studies to illuminate such connections. And of course, with every year that passes come other experiences and exposures that contribute to the onset of illness, making direct links to prenatal life hard to untangle. But the groundwork that the nine months of gestation, and the first years of childhood, lay for the sickness and health of a lifetime is a burgeoning area of research. Scientists are starting to ask whether airborne toxins might wield their power in slow motion, creating a hidden susceptibility that lies dormant for decades, until it is touched by some fresh exposure or trauma. Even those of us apparently unaffected by what we've breathed, in other words, may not have escaped the consequences. We just haven't felt them yet.

I tell Ritz how surprised I've been by the sheer range of illnesses linked to dirty air, the terrifying variety of the marks it leaves on us. "It's not surprising to me," she replies. "If you see it all as independent effects on different diseases," then each one is a new, and shocking, piece of news. But if you look at the bigger picture, the interconnected nature of the human body and all its systems, it makes more sense. "We are a whole. Only we have weaknesses. And I guess the disease we get is the one where our weakness kind of overwhelms our defenses."

* * *

Ed Avol and Jim Gauderman's pioneering research didn't end with their first set of findings, the breakthroughs on lung development. They're still following the threads they began teasing out decades ago, and they've agreed to tell me about the journey in person. So one sunny January afternoon, I walk past office doors covered in charts and graphs, into a bright, sleek conference room at USC's Keck School of Medicine. Avol, a wiry man with the intensity of a long-distance runner, leans forward as he talks, words bubbling forth. Gauderman speaks more slowly, with a smile that's broad and easy beneath his trim moustache. They're friendly and earnest, straightlaced science guys and confessed geeks, although when Avol proffers the label, Gauderman genially objects. "Speak for yourself," he laughs.

Over the years, the Children's Health Study has expanded far beyond its original driving question to address dozens of new ones, grounded, always, in the bedrock of its extraordinary data set, those detailed portraits of thousands of children's formative years. That trove grew more granular as the researchers scraped DNA samples from subjects' cheeks, distributed step counters to monitor physical activity, and tested the children's breath for a chemical that warns of inflammation.

The study "has gone off in all kinds of directions," Avol says. "We sort of grew from just looking at respiratory health, which was our original interest, to looking at cardiovascular health, looking at neurological health, looking at obesity and metabolic syndrome, looking at genetic predispositions, looking at epigenetics." It has drawn in experts from a multitude of disciplines—geographical mapping, toxicology, molecular biology—even teams examining mice to trace pollution's pathways through living bodies.

It's that intellectual richness, the interplay of so many disciplines, that has kept the men engaged for all these years. In reality, Gauderman explains, the study is not one project, but many. "That's the key. It started out as twelve towns and 'Does air pollution affect chronic health in kids?'" That was important enough, but the constant lure of new questions has made it more compelling still.

Even more gratifying is the awareness that their work is making a difference. The early findings landed on the desks of people in Sacramento and Washington with the power to change the laws and rules that determine what's in the air breathed by millions. And those officials took heed—not always, but often enough to matter—using the science to demand that cars, trucks, factories, and power plants get cleaner.

So now, Avol and Gauderman are in the extraordinary position of studying health benefits their own research helped bring about. Over the years, as Southern California's air cleared, Gauderman started sifting again through the old data, comparing Erika Fields's generation with kids growing up in the new millennium. This time, the results were less surprising: The younger cohort, breathing better air, is far healthier—the rate of serious lung function loss is less than half what it was.[34] When I point out that the Children's Health Study had a lot to do with that change, they acknowledge their role, modestly. "We're in a department of preventive medicine," Gauderman says, before his partner finishes the thought: "We used to have a chair who used to say, 'Go out and prevent something.' So we like to think we contributed."

Of course, the work is personal, since their own lungs were developing back when smog alerts and warnings to stay indoors were as much a part of Southern Californian life as orange groves and beaches.

Avol cracks a corny joke, warning me to take Gauderman's words with a grain of salt because "these things do have cognitive effects." But they know better than anyone how serious this is. Gauderman thinks about his own kids' health. They've grown up with far cleaner air than he did, and he and his wife chose a home far from busy highways. But it's too soon to know what future illness the air of their childhood may have already, invisibly, set in train.

It was William Wordsworth who wrote that "the child is father of the man." Today, we don't like to think our path through life is set so early. Modern parenthood, though, is filled with worry about how our decisions—schools, screen time, sugar—and even our words, may affect our children decades from now. Did we encourage them enough to build confidence, without undermining independence? Work enough to provide a secure home and a good example, but not so much that we weren't there when they needed us? I know my nightly argument with Anna about whether mac and cheese with cookies is a balanced meal may affect her relationship with food as an adult, just as my parents' penchant for stashing sweets in a hiding place whose location we all knew probably underlies the forbidden allure chocolate still holds for me.

Child-rearing is imbued with awareness that the nurture we provide will have repercussions later. Yet few of us realize that the air our kids breathe—the air we've chosen to expose them to, or that has been chosen for us, by the circumstances of jobs, of money, of family obligations, that dictate where we live—may also be shaping their futures. A saying sometimes attributed to the Jesuits phrased Wordsworth's sentiment a little differently: "Give me the child until he is seven and I will show you the man." What consequences, I wonder, will the diesel fumes of my daughter's London childhood hold for her when she is a woman?

Avol and Gauderman know the change their work has helped spur is not enough. So the team is still pushing forward. Today, much of their time is spent on a question that eats at me every day: How much damage is caused not by the clouds of pollution that sit over a whole city or region, but by the especially high levels right beside roads? Rob McConnell is part of the Children's Health Study too, and when he joins us in the conference room, he laughs at how sure he once was that such variations couldn't possibly matter. Early on, he went to Chile with a colleague. "We spent a whole afternoon with traffic engineers," McConnell recalls. "I remember thinking, 'What is this? Why would he think that little blip that you get from near-roadway exposure would have any effect?'"

It turns out the effect is huge. Pollution concentrations can be two or three times higher by a busy roadside than even a hundred yards away.[35] So, while the first phase of the Children's Health Study had compared kids in clean-air communities with those in polluted ones, in later years the scientists have drilled deeper, examining differences within each community, zooming in to look at how air quality varied from one block to the next and trying to tease out the effects of living on a busy street from the impact of being in a town or neighborhood where average air quality is poor. They tested at children's schools and homes, and along the roads they walked, and Gauderman created an individualized picture of the air every single kid breathed over years.

My heart sinks as he explains that the effects are additive, that together, poor overall air quality and proximity to heavy traffic deliver a double whammy of trouble. Because that, unfortunately, is a pretty good description of my life in London, where the citywide pollution is high and our house, while on a quiet, dead-end street, is just a short walk from a road where cars sit bumper to bumper every rush hour. At her school on a road traversed all day by hulking trucks and double-decker buses, Anna's lungs are likely getting an even bigger dose of exhaust. Spikes like that, on and near the busy streets where so many of us spend much of our time—strolling to work, driving, sitting in our living rooms—make pollution a threat even in places where overall air quality is good.

As afternoon turns to evening and a pickup basketball game heats up outside the conference room, McConnell tells me about the Colorado hospital where his mom was treated after a heart attack. It sat beside a major highway, and he couldn't help thinking when he visited about the evidence suggesting air pollution causes arrhythmias, clotting problems, and other changes dangerous for heart patients. Even putting the parking lot between the road and the hospital would have made a difference, he says.

The building's designers probably didn't know that, but zoning officials should, and they can make rules to reduce unnecessary exposure. "If you're building a new school, why would you build it next to a freeway?" he asks. Exercise greatly increases the amount of air—and thus, the pollution—our lungs take in, so McConnell wishes the runners he sees along L.A.'s Sunset Boulevard knew how much better off they'd be on one of the quieter roads that parallels it. Those who do, he believes, ought to nudge them in that direction.

Another realization Rob McConnell has had is that whatever we do now, pollution's toll is likely to grow. For he found something in his data that had never occurred to him, but seemed utterly obvious once he discovered it: "We're getting older." The effects of decades of breathing dirty air may not be apparent in the prime of life. It's later, as the damage snowballs and the body weakens, that the strokes, the heart attacks, the cancers do their worst. "It shouldn't have been surprising, but it was," McConnell says. "You don't see the deaths in young people. Where you see the deaths is in old people." As we begin to pay the price for what we inhaled long ago, he says, pollution-driven illness and health costs will skyrocket, even if the air gets cleaner.

The frontiers of this research keep moving forward. Now, McConnell is looking at dirty air's relationship to obesity and diabetes, conditions that, even a decade ago, would never have been part of a pollution study. One day, he bumped into a former department chairman in the elevator. "What's with you guys?" his old boss asked. "You think air pollution causes everything?" On some level, McConnell says, it's not an absurd contention. He draws an analogy to smoking. No one is surprised to learn cigarettes cause damage far beyond the lungs. While the mix of toxins coming from cars and factories is different from the stew in tobacco smoke, the principle, he tells me, is the same. The molecules we draw in with each breath touch every corner of our bodies.

* * *

John Miller understood the science of air pollution, too, had read the papers, digested the data. But none of it prepared him for the day a boy with a stopped heart was rushed into his emergency room. Miller speaks with the slow drawl of his native Tennessee, but the hospital was in Orange County, the densely populated sprawl of suburbs just south of Los Angeles, not far from where Avol and Gauderman spend their days.

The call came during what had been, until then, a quiet shift: paramedics were on their way with a child in cardiac arrest—a patient, in Miller's recollection, around eight years old. "In an emergency room, the day can change in just instants," the doctor tells me. He scrambled to marshal the resources he'd need when the ambulance arrived. Within moments, a pediatric crash cart was at the ready. So were the tubes and intravenous lines that would sustain the boy.

It was during that interval, too, that someone issued an ID number and printed it on labels that would be used to mark blood samples and other specimens for the lab. The hospital didn't yet know the boy's name, so the tags just said "John Doe."

They could be updated later. For now, all that mattered was restarting that stilled heart. At times like this, "the adrenaline's flowing," Miller says. "You're just in hand-to-hand combat with death and destruction." He and his team did CPR as the terrified family waited. "Then, all of a sudden, we got a heartbeat back, and he started breathing on his own," the doctor continues. "We got a blood pressure. We've got a patient who's going to survive, and most likely survive well."

Air quality was particularly poor that morning, and the child's crisis had been triggered by a severe asthma attack. When his mother said he often wheezed on smoggy days, the cause of this trauma became clear to Miller. That may be why it stayed with him while the stories of so many other patients, even of children who weren't as lucky as the one he saved, have not. "It just seemed so obvious to me," he says now. "The only possible explanation for that disaster was the air pollution we had that day."

He'd seen dirty air's effects routinely at the hospital, where smoggy days predictably meant a rush of patients struggling to breathe, asthmatic preschoolers on nebulizers and elders with oxygen tanks. Always, at the back of Miller's mind, was the knowledge that, whatever happened to a patient, the job of walking down the hall to tell the family would fall to him.

As it happens, air is something Miller knows a lot about. Outside the ER, he'd testified at Senate hearings, sat through endless municipal meetings, and fought for changes at the L.A. area's two big ports, which are responsible for a big chunk of the region's pollution. When I meet him one morning at a fellow activist's home, Miller hands me a paper from the *Journal of the American Medical Association* showing death rates climb inexorably when particle pollution rises. "That, to me, was stunning," he says. "If we have succeeded in decreasing the air pollution burden by even a small few micrograms per cubic meter, I will have saved more lives than I ever did working as an ER doctor."

But while he knew pollution's power, the fight to snatch one child from death's grasp drove home for Miller in an entirely different way the reality that "there's actual people that get put in body bags because of this stuff. He wasn't one, thank God. I was really glad that we saved his life—I guess I was proud of it, too, to be honest with you, proud to save him. But thinking it was foolish that I should be having to do this," that a man-made hazard causes so much needless suffering.

After the boy and his relieved parents went home, Miller slipped one of the John Doe name tags into his wallet. As it grew threadbare and tattered among the cards and cash there, he pressed on with his activism, with the hearings and the interviews and the relentless pushing and fighting. When things got discouraging, when a decision went the wrong way or a company reneged on a promise to clean up, he'd pull out the label "to remind myself to try to speak out for people who could not speak out for themselves," he says. "Sometimes I would think about, 'Why am I doing this? Why am I beating my head against the wall on this thing?'" The ports and the trucking firms and the other polluters may have had more money and more lawyers than the communities around them could ever afford. But the faded ID, and the memory of the child it was made for, kept him focused on exactly what was at stake.

It's been about a decade since Miller coaxed that young heart back to life. Somewhere in the world, its lucky owner is a teenager now, nearly a man, old enough to shave, maybe heading to college. He may not even remember the day he nearly died or the doctor who made sure he didn't, although his parents will surely never forget. Miller has since retired from medicine, and the child's name has disappeared into the mists of his memory. But seared there forever is the knowledge of what air pollution did to him.

The boy's case was unusual not for its drama, or even its severity, but for the clarity with which its cause could be pinpointed. He stands, in a sense, for all those who endure disease and loss because of the air's foul taint, but who are unable to make out the true culprit. One child whose mercifully brief trip to the very threshold of death pierced a doctor's professional armor, and reminded him what all those statistics and studies are really about.

2

GROUND ZERO

Delhi's Health Emergency

PM2.5. Letters, and a number. The dry label disguising a mortal threat. Particulate matter, smaller than 2.5 microns across. Particles half the size of most bacteria, smaller than some viruses. A thirtieth as wide as a hair. Able, together, to dim the sun, blur buildings and mountains. But visible on their own only with the most powerful of tools. Under the gaze of an electron microscope, they are dark and menacing. Each one different: some smooth and round. Others jagged and many-sided, stretching into irregular ovals, into bent chains and branches. Or squashed, like a misshapen hunk of clay, a bit of dense fluff. Made from all the elements in all the things humans can think to burn: carbon and silicon, iron and aluminum, titanium and sulfur. Copper, tungsten, lead. Sometimes changing shape, distorting themselves to grasp a deadly new partner. Still light enough to travel on the wind, thousands of miles, across oceans and continents. To sneak through the crack beneath a window. To go anywhere the air goes. Until someone, somewhere, breathes it in. Into the body now, down through the lung, deeper and deeper.[1]

To study religion, the devout go to Jerusalem. Lovers of art make their pilgrimages to Paris. I'm a different sort of tourist. I want to understand what it means to live with truly abysmal air, to breathe in a place whose pollution is among the worst in the world. So I've come to Delhi.

India is often said to simultaneously enchant and overwhelm Western visitors, the delights of its food, its temples, and its vivid colors competing for mental real estate with a crushing sense of the sheer number of people and a soul-tugging awareness of the grinding poverty that afflicts millions. The clichés all feel true—I am duly enchanted and overwhelmed—but it is my awareness of the public health emergency unfolding around me that colors my time here even more sharply.

As I traverse a city where homeless toddlers play beside traffic and lush private gardens hide just off hectic thoroughfares, I simply cannot shake the knowledge that 1.6 million Indians die every year[2] because their government can't or won't stop the burning—of plastic and other garbage, of diesel, of cow dung, of the cheapest coal and the stalks left over after rice and wheat are harvested—that makes this country's air so catastrophically bad.

Descending into Indira Gandhi Airport, I watched its lights through the heavy haze, and understood before the plane touched down that the air is no abstraction in Delhi. Even in April, one of the cleaner months, I can see pollution shimmering in the streetlights at night, like the wispy smoke that floats from a campfire. By day, it's a gauzy curtain hiding bridges and buildings.

This place—Delhi in particular, India more widely, and South Asia as a whole—is air pollution's ground zero, the epicenter of a health crisis of stunning proportions. In megacities across the subcontinent—Hyderabad and Kolkata, Karachi and Dhaka—hundreds of millions of people are assaulted with each breath by toxins from a mind-boggling multitude of barely regulated sources. India is home to 9 of the top 10 cities on the World Health Organization's list of the most polluted. While China's terrible air makes global headlines, Delhi's average PM2.5 levels are nearly double Beijing's, more than a dozen times the recommended maximum, and exponentially worse than in even the most polluted Western nations.[3] Sometimes, during awful, days-long stretches when thick clouds of smoke settle in over the city, particle concentrations spike so high that schools close in a desperate effort to protect pupils, and cars collide because drivers can't see stoplights. Hobbled by corruption and overwhelmed by the many problems that come with deep and widespread poverty, India's government seems utterly inadequate to the task of protecting its people, far weaker than a Chinese state practiced at imposing its authority.

The human consequences are dire: One study[4] found Delhi's school-children are twice as likely to suffer serious lower respiratory problems like pneumonia as a control group in a less polluted area, and they have a 65 percent higher incidence of poor lung function—that irreversible weakness, so predictive of serious health problems later in life, that Ed Avol and Jim Gauderman found, in far lower numbers, in California's children. Nationally, pollution ranks behind only malnutrition as a driver of death and illness.[5] Indians have the worst lung function in the world[6] and the highest rate of death from respiratory ailments.[7] In most of the world, chronic obstructive pulmonary disease—a severe collection of breathing problems characterized by scarring, inflammation, and narrowing of the airways—is mainly caused by tobacco. But in India, some 60 percent of those diagnosed are nonsmokers.[8]

We'll visit wealthy Western nations in this book, too, and their air quality problems are real. But one terrible figure offers a stark illustration of how much more severe the danger is for places like India: Children under five are 60 times more likely to die from the effects of dirty air in poor countries than in rich ones.[9] Whenever I speak to Indians with the means to travel abroad, they tell me about health woes that vanish the moment they leave. A child's wheeze, a father's shortness of breath, a nagging cough—all disappear on visits to Europe, Canada, the United States, and reappear almost instantly upon arrival back home.

"You feel like there's like an ulcer or something inside you," said Manjali Khosla, who returned to live in Delhi after decades overseas. "It's like some burning sensation, and then the sinuses hurt, your head hurts, your eyes water." Her children hold their noses when they get off the plane after vacations in Canada, and their mother is terrified of the longer-term consequences. "I've even dreamed polyps coming out of their lungs, something very drastic," she tells me.[10] Delhi's residents could only laugh at the dark irony when, just after his election, their chief minister, the equivalent of a mayor, left town on his doctor's orders in hopes of shaking a chronic cough.[11]

* * *

Here's one way to cope: Kamal Meattle has built himself a bubble. The Paharpur Business Center is a hushed, almost otherworldly, escape, its cool cleanliness a welcome contrast to the sweaty cacophony of Delhi. From the outside, the building is unremarkable, a slightly grubby, 1970s-style white office tower squeezed beside a big, dusty parking lot, a fancy facade of black stone and dark wood looking a bit out of place around its front door. Inside, the decor, replete with more polished wood and lots of upholstery, looks like that of a home whose owners care for it lovingly but haven't bothered to keep up with changing tastes. There's not much natural light, and the long, dark corridors and windowless rooms give the place a womblike feeling. It's not Paharpur's appearance that sets it apart, though, but the elaborate mix of high- and low-tech defenses that make its air some of the city's most breathable.

Big multinationals rent office space and conference rooms here, but Meattle really created Paharpur as a haven for himself, a stage on which to play out his obsession with air. The idea came after his doctors told him in the mid-1990s that if he stayed in Delhi, its pollution would soon kill him. His lung capacity had fallen precipitously, he coughed all the time, and he suffered hearing loss from the allergy pills he needed to control his symptoms. He decided to defy the advice and stay, asking for help from scientists at the Indian Institute of Technology Delhi, whose board he served on.

They came up with a greenery-based approach to air purification— Meattle says it was inspired by work at NASA—so 1,200 plants crowd the building. They line every hallway and jam a rooftop greenhouse, where they grow in big barrels on the floor and peek from plastic water bottles strung up along the walls. For a while, the plants alone were enough to make Paharpur's air pristine, but as Delhi's pollution levels soared, bigger guns were needed. Now, outside air is sucked into a tank on the roof, where it is washed and filtered, then pumped through the greenhouse to boost oxygen content. An operating room–style positive pressure system prevents contaminants from coming in when doors are opened.

Meattle, who made his fortune in packaging, is deeply detail oriented, the kind of guy who tells you about the patents he holds and boasts of the TED talk he gave. As we speak, he barks instructions to an assistant projecting slides and videos on a screen at the front of the room. "Just open that presentation," he orders. "The plants, show the plants also. Slide by slide, just run through it." He forbids workers from eating anywhere other than the cafeteria for fear vapors from their food might taint the building's air. When there's no rain, he has the 40- and 50-foot-high trees outside his home washed every few weeks so pollution won't interfere with photosynthesis. He rattles off topics I should Google, suggests one he thinks could form the basis of a PhD thesis, and tosses out oddball ideas, like the suggestion that countries with the cleanest air could offer blood oxygen tests on arrival and departure so visitors could see how much a stay has improved their health. "That's a new idea," he says. "But anyway, that's on the side."

The bigger point, he says, is Paharpur's effect on well-being. Undoubtedly, the building offers a pleasant respite, and Meattle points to a study showing headaches, breathing troubles, and eye irritation are much lower than in other offices. His TED talk was titled "How to Grow Clean Air," and he says the oxygen the plants give off boosts workers' productivity. It's not long before his aide clips an oxygen monitor to my index finger. Meattle assures me my reading of 98 percent indicates an acceptable level of respiratory health, although "chances are you'd become 99" after a day in the building. "I would love to see what happens to blood oxygen levels of all the people living in Delhi versus people living next to the beach somewhere," he muses. His plants, Meattle tells me, would be useful to anyone sheltering after a nuclear attack. If you pumped in outdoor air to replenish oxygen in such a situation, "you'd be all dead. I said, 'Great, this will have a defense application, let's work towards it,' and we've been able to do it." A faint impish smile comes over his face, and I think he may be joking, but it is hard to be sure.

Meattle is healthy now, but he knows his building does not solve the bigger problem. Barun Aggarwal, his son-in-law and business partner, agrees air filters are little more than a bandage on a gaping wound, but he's started selling them for home and office use. "The government needs to do something, and the sooner, the better," says Aggarwal, the American twang in his voice evidence of 13 years spent in North

Carolina and Atlanta. But he's not waiting. "Am I just going to let my kids breathe that poison for the next 5 years till the government gets their act together? No. If I can put in an air purifier, give their lungs some rest, get their wheezing down, get their coughing down, give them some clean air, I will do that."

Aggarwal, a longtime entrepreneur, didn't think much about air for the first couple of years after he returned to India. He was traveling often to Atlanta, training for a Thanksgiving marathon there. Late one night in Delhi, he went for a two-and-a-half-hour run. He recalls "coughing black sputum out for the first 30 minutes after coming back. I couldn't believe it." On trips to Atlanta, he would train with a friend: "We'd just start the run and I'd just go 'Aaahh, I feel so good!' And he'd look at me, like, 'What's wrong with you? Why are you saying that?'" Aggarwal would try to describe how awful the air was at home, but his friend dismissed it. Soon after, the same running buddy visited Delhi, and Aggarwal took him for a jaunt near India Gate, in the city center. "Five minutes into the run he said, 'I can't do this anymore, let's go back,'" Aggarwal recalls. "He says, 'Now, I know what you mean.'"

Aggarwal shares his father-in-law's obsession with air, carrying a high-tech PM2.5 monitor to measure particulate levels everywhere he goes. One autumn, he visited Rishikesh, the hill town where the Beatles once sought enlightenment. "I got a reading of 21 micrograms per cubic meter. Same day, I came back to Delhi: 489." The WHO guideline says a city's annual average fine particle level should not exceed 10 micrograms per cubic meter; allowing for occasional spikes, any one day's average can reach 25. Aggarwal's highest-ever reading came during the festival of Diwali, which Hindus celebrate with ear-shattering fusillades of fireworks. Outside his home, in one of the city's greener areas, he says, "I got a reading of 7,200 micrograms per cubic meter. The visibility, I couldn't even see one meter." Breathing in Delhi on those holy nights feels "like lighting a fire and putting your face in the smoke."

Aggarwal shows me a short video made by Josh Apte, an air expert from the University of Texas at Austin. On a visit to Delhi, Apte brought his monitor for a ride in an auto-rickshaw. The green-and-yellow three-wheelers are everywhere in the city, a quick and afford-able, if sometimes harrowing, form of transportation. They are door-less, with open sides, and drivers sit at motorcycle-like handlebars in front, while passengers perched on the seat behind hold tight around curves and corners. Everyone on board inhales stunning amounts of pollution.

Aggarwal narrates as the video flashes in front of us, the parti-cle level displayed atop the screen as the auto-rickshaw speeds, then crawls, through traffic. The PM2.5 concentration jumps around: 545, then 335, then 895, dozens of times the safe limit. As a noisy truck grinds its gears in the next lane, the reading climbs to 1,427. "Oh, my God," I gasp. Aggarwal laughs: "Just you wait." After a moment, a heavy plume of black smoke appears out of nowhere, billowing toward the camera. "Now watch," he says, and the particle readout spikes to 2,621. It never goes lower than 159, more than six times the WHO's recommended maximum. It's no surprise Apte declared pollution in Delhi's auto-rickshaws "among the highest values ever reported for a routine transportation microenvironment."[12] This, says Aggarwal, is the true measure of what Delhi's 16 million people breathe, "the real exposure at the ground level."

* * *

Few of those millions will ever get a gulp of Paharpur's rarified air. India is a country still riven by class, and while air pollution tran-scends socioeconomic divides, affecting everyone with a set of lungs, there is no doubt that the rich, the poor, and those in the middle all feel it differently. Traffic exhaust is just one of many sources of Delhi's pollution, but the long hours many ordinary people spend on and beside the streets each day are a big reason for the disparate impact.

A few miles from Paharpur, on busy Mathura Road in south Delhi, a woman's smiling face beams down from a huge billboard advertis-ing free Skype calls. "Green Delhi Clean Delhi," boasts a dirty white banner that bends around the corner just beneath it. Although it's Sunday, traffic is near a standstill. Sleek SUVs and packed buses inch along, drivers leaning into their horns. Beside the billboard, a vendor carries a pile of wooden recorders in a bundle tied to a pole, and the high-pitched whistle he blows adds to the ear-splitting noise.

Conversation seems almost impossible, yet the small patch of sidewalk bursts with activity, a hub for cooking, eating, and the idle passing of time. It's late afternoon, and families are heading home. Men in turbans and women in brightly colored saris push past, dragging boys with neatly combed hair and pressed shirts. A little girl in a dress covered with gold threading reaches a hand up to touch the Green Delhi sign as her brother, in a smart white coat festooned with bright blue ribbon, stands nearby. Just next to the billboard, in a row of food stalls, two men bake chapatis, and their customers scoop up lentils with pieces of the round bread. A few feet away, another vendor pours thick lassi, a delicious yogurt drink, from a metal jug as incense smoke billows from his cart, adding a heavy, fragrant note to the acrid diesel smell that wafts from the road.

The corner seems to be a popular spot for auto-rickshaw drivers looking for fares; a half dozen or more congregate beneath the big signs. How do these men, in their worn gray uniforms, view the pollution that is an inescapable part of their working lives? Neha Tara Mehta, a Delhi journalist just back from a stint living in New York, is helping with my research here, and she begins buttonholing drivers to find out. Unlike educated Indians, few of the poorest speak English, but Neha translates my questions into Hindi, and the drivers' frustration quickly bubbles out. Many tell us they began life in small villages where opportunity was scarce, and came to the capital as young men, hoping for steady work. Years later, they struggle to make ends meet, and the brutal days they spend in the heat, noise, and exhaust take a painful toll.

"We are the worst affected because we drive an open vehicle," says Nand Lal Kumar. "But nobody thinks about us. Nobody cares about us or our families." Angry and animated, Kumar tells us that after 17 years in this job, he barely earns enough to send his children to school. "I can't even pay my rent on time," he says, a tattered white T-shirt peeking out from beneath his uniform. On the road, conditions are hard to bear: his head pounds, and he's wracked by coughing. "I feel suffocated, anxious," he says. "My chest gets heavy, stomach gets tight." He wipes soot from his face at the end of the day, and sees a respiratory specialist for a chronic cough. "'Take some medicine and you'll be fine,' that's what the doctor says. It gets OK and then it gets bad again." He'd love to do something else, but "there's absolutely no other alternative. There are no other jobs." What will come next for Kumar? He doesn't know. "I do worry about my future health, but there's nothing I can do," he says. "My job is a hazard for me."[13]

* * *

For all the drivers' suffering, their auto-rickshaws are among the few vehicles that don't contribute much to the bigger problem. It's hard to believe now, but not long ago, Delhi was seen as an air pollution success story, a model of how a vast developing-world megalopolis could clean its air. The auto-rickshaws played a part: in 1998, the Supreme Court ordered them to convert to cleaner fuel, mostly compressed natural gas, known as CNG.

"Let me take you back in time a little," says Anumita Roychowdhury, an elegant woman in a beige and pale blue wrap. She's the director of the Center for Science and Environment, a group that's played a leading role in the years of battles over air quality. In the 1990s, she tells me, Delhi's air was so bad "you couldn't go out in the city without your eyes watering." India had no regulations on vehicles or fuel, so despite advances elsewhere in the world, engines here hadn't improved for 40 years, and fuel quality was abysmal.

It was the activist Supreme Court that changed that. Its judges started issuing orders, and from 1998 to about 2003, a series of important new rules came into force. Polluting industries were pushed out of the city, auto-rickshaws and buses were converted to CNG, and emission limits for vehicles were introduced, then tightened. "These were pretty big steps," Roychowdhury says, and they brought results. "If you plot the graph of particulate matter in Delhi, you will see after 2002 the levels actually coming down." The public noticed. "I still remember the 2004 Assembly elections in Delhi, where the political parties were actually fighting with each other to take credit for the cleaner air. It had become an electoral issue."

So how did things go so wrong? The burst of activity petered out, and rapid growth in car ownership erased the improvements that had been won. "If you look at the pollution levels again from 2008 and '09 onwards, you now see a steady increase," Roychowdhury says. "We could not keep the momentum going." Indeed, particulate levels jumped 75 percent in just a few years.[14] Even the action that was taken, she believes, "was too little. We had to do a lot more, more aggressively." Part of the reason government stopped pushing, Roychowdhury believes, is that the moves needed next would have had to address Delhiites' growing fondness for cars, so would surely have prompted public anger. "There is a hidden subsidy for all of us who use cars today," she says. "We barely pay anything in terms of parking charges, we barely pay anything in terms of road taxes. It is so easy to buy a car because of easy loans. So there is absolutely no disincentive." About 80 percent of transportation spending is focused on drivers, even though they're only about 15 percent of Delhiites. "The entire infrastructure of the city is getting redesigned to facilitate car movement, but not people's movement."

In fact, there have been big public transportation improvements; they're just not enough. Delhi's Metro began running in 2002, and the system has expanded steadily since. Creating such a network from scratch is a real achievement, and traffic would surely be even worse without it. But mass transit remains badly inadequate, and a lack of big-picture planning makes it hard to use. There is a serious shortage of buses, for example, so many passengers find themselves unable to get to and from Metro stations.

Consequently, the number of cars and motorcycles in the city climbs ever upward, and Delhi's streets grow more clogged by the day. Drivers are a must-have for the upper-middle class and the wealthy. On my first workday in Delhi, Amit Kumar appeared smiling at my guesthouse moments after I requested an all-day taxi, and he chauffeurs me around throughout my stay, dropping me wherever I need to be. At $18 for an 8-hour day, it's a nice way to travel, especially as temperatures soar above 100 degrees. More than that, I honestly have no idea how I would function here without someone to drive me around.

Many locals, it turns out, feel the same. "You can triple the size of the Metro, but I don't think you will get Samir in the Metro, it just won't happen," one think tank analyst says of his colleague. "Or you won't get me." Having a driver is a status symbol, analyst Vikrom Mathur tells me. When he lived in London, he rode the Underground and the bus regularly. "I have been back in Delhi for three years, and I have been in public transport twice."

Every city has traffic, though, and activists say the air in Delhi doesn't need to be so much worse than elsewhere. Emission rules for new vehicles are being tightened gradually, but lag far behind even those of many other developing nations. Environmentalists' demands for a faster ramp-up have brought years of battles with automakers.

But tightening requirements on new cars doesn't address a more immediate issue. Vehicles sold before stricter rules kick in—including ancient clunkers that predate the introduction of even the most basic pollution standards—can remain on the road indefinitely. "In a country like India, where we don't throw away anything, there are 15-year-old, 20-year-old vehicles on the roads," an auto industry spokesman tells me.[15] He's right. Many cars are driven for far longer than the effective life span of the catalytic converters that detoxify their exhaust, making a mockery of any standard they might purport to meet.

It's not just cars. The sheer number and variety of sources feeding into the foul brew that Delhi's residents must breathe is mind-boggling. Mom-and-pop-scale industrial activity dots roadsides everywhere— gray smoke pouring from a metal barrel next to a makeshift food stand, sparks flying as welders work in a wooden shed. Diesel generators power many cell phone towers and provide backup electricity for offices, hospitals, and homes during frequent blackouts. Construction sites create dust that contains asbestos, silica, and other hazardous materials; it endangers laborers and spreads through neighborhoods.

Pollution from beyond the city limits adds to the mess. In the nearby states of Punjab and Haryana, rice and wheat farmers torch their fields every autumn after harvest. The smoke that drifts to Delhi is so thick NASA photographs it from space: on the images, red dots, representing fires, cover the area northwest of the capital, and clouds of smoke engulf the city. As predictably as the turning of a calendar's pages, the capital endures eye-popping particulate spikes every November, but authorities haven't yet bothered to help the barely surviving farmers find a better way of clearing their fields.

What's more, 700 million Indians, mostly in rural areas, burn dung and wood to cook. Until recently, scientists believed the health-destroying effects of their smoke were mostly felt close to home, but now say it floats toward cities and accounts for more than a quarter of the country's overall air problem. Coal-fired power stations are another big contributor, causing about 100,000 premature deaths and 900,000 emergency room visits nationwide every year.[16] Regulation is lax at best, nonexistent at worst, so the plants almost always lack the scrubbing systems required in wealthier nations, and the smoke wafting from them is laden with toxins.

Which is the biggest contributor? It's hard to say. Elsewhere, "source apportionment," an inventory of the causes of a region's air pollution, provides a framework for cleanup efforts. In Delhi, such rigorous studies were, until recently, hard to come by. Even the numbers available now are sometimes contradictory, and often angrily disputed. One government-commissioned report blamed dust from unpaved roads for more than half the problem and said vehicles contributed just 6.6 percent.[17] Automakers embraced that estimate, while environmentalists ridiculed it, accusing officials of trying to downplay traffic's role. One independent study put vehicles' contribution at just over half, another said 20 percent.[18] What is clear is that the sheer number and variety of pollutants adding to the foul mix in Delhi's air—and India's—make the task of cleaning up more daunting than just about anywhere else.

*   *   *

The air experts I meet keep mentioning brick kilns, a polluting industry I haven't come across elsewhere. So I ask Amit, my easygoing, resourceful driver, if he can show me one. He speaks to a friend, who speaks to a friend, and before long we're bouncing along a narrow, bone-jarring road just outside the city, translator Neha sitting beside me. Amit parks on a patch of dirt, and several dust-covered men amble over to greet us. We follow them up a small hill, where a gray smokestack, its top blackened by soot, towers over a brick hut. Surrounding both structures is a large, U-shaped canal, some 15 feet deep and 25 feet across. It runs about 70 yards from end to end, before turning a corner and stretching back the same distance on the far side of the chimney. At one end of the U, bricks of pale beige clay are stacked neatly, reaching from the floor of the canal up to ground level.

I stop for a moment, trying to understand what happens here, but my attention is pulled farther along, to the wide, dusty fields stretching for a quarter of a mile or more beneath the low mound we're standing on. A hundred or so people are scattered around them, squatting on the ground or wielding shovels, cloths wrapped over their heads in the baking sun. This is where the raw bricks come from, and as the manager leads us along, I wish I had worn something sturdier than sandals on my feet, which are already coated with a thick layer of dirt. Up ahead, a group of boys, eight years old or so, ride bikes along paths cutting through the fields. Children are everywhere, playing as their parents work and helping where they can. A small girl, barely more than a toddler, lugs an even smaller one up a gentle incline. In the distance, I can see the chimneys of a half-dozen kilns similar to this one, heavy black smoke floating from those where firing has begun.

The ground is covered with bricks, beige like those I saw in the kiln, dried by the sun but not yet fired. They stretch out in long lines in the dirt and are piled high in makeshift walls. We head toward a woman who squats beside a thousand or more of them, laid out in rows. Moving quickly but precisely, she scoops mud into a brick-shaped mold, scrapes off the excess, and places the finished product with the others to dry. Her sons keep her supplied: the 15-year-old shovels mud into a wheelbarrow, and his younger brother places it in large patties on the ground. Together, they make 2,000 bricks a day. With supervisors listening, the woman, whose name is Amarwati, doesn't complain—"I just keep sitting here, working," she says—but the job is clearly backbreaking. I learn later that the millions of workers at the 150,000 or more kilns around India endure horrific conditions. Some report having their hands chopped off, and unscrupulous payment practices leave many trapped by the shackles of debt that can be tantamount to slavery.[19]

After we've explored the fields, the foreman leads me, Amit, and Neha back toward the tall smokestack and walks us into the empty firing canal. With the high walls stretching up around us, he explains how it works: The canal holds nearly a million bricks, with coal placed into gaps between stacks. One end is lit, and as the fire moves along the U, finished bricks are taken out and replaced with new ones before the burning comes back around, a process that goes on six or seven months a year. Vents in the walls route smoke to the chimney, and I ask whether it is filtered before floating away. "No filters," answers the supervisor. The bricks are sold around the region. "All these new buildings are rising because of us," he boasts, gesturing to a nearby apartment tower, and indeed, the kilns have supplied a huge construction boom.

He dismisses the idea that his industry damages Delhiites' health, but the experts say otherwise. Although the kilns are not allowed within the city's borders, about a thousand of them sit just outside; they may be responsible for as much as 10 percent of its particle pollution.[20] And it's not just the capital's problem; kilns surround many urban areas. It doesn't have to be like this. There are better ways to make bricks. But operators feel little pressure to change.

A few minutes after leaving the kiln, we pass a sign welcoming us back into the city. On a narrow road crowded with speeding cars and motorcycles, smoke billows into traffic, so thick it's hard to see. Drivers zoom on, barely noticing this toxic cloud. Garbage fires are routine in Delhi, a way of getting rid of the rubbish that piles up everywhere. Here, an open field is strewn with trash, and a nearby shop owner tells me someone lights it once or twice a month, whenever the pile grows too high. This time, the fire has been going for an hour or two. "There's a lot of garbage, God knows how long it will keep burning," he says. The sun is getting low, and the scene is strangely beautiful, the light glowing red through clouds of smoke. Women in bright colors walk past, dodging traffic and pulling scarves over their mouths and noses. There is lots of plastic in the pile, and the rancid smoke is stinging.

This is a busy commercial neighborhood, its main street lined with open-front shops, and there must be thousands of people within close range of the fire. It may seem obvious, but scientists say it's that proximity that makes sources like this one so deadly. The latest thinking on pollution's effects focuses less on what is emitted than on what people actually breathe. While coal-fired power plants, for example, are big polluters, only a small fraction of what they spew ends up in someone's lungs. On the other hand, a good deal of the exhaust from cars in densely populated areas finds its way into human bodies. The same goes for smoke from the fires that security guards kindle outside the homes of Delhi's wealthy to keep warm in the chilly winters, or those homeless families build from trash or sticks to cook.

It's also true of fumes from the hulking trucks that rumble through Delhi, an estimated 80,000 of them every night, their cabs painted in bright hues, bumpers and windshields plastered with gaudily colored decorations. Barred from the streets until well after dark, many are intercity haulers traveling from one end of India to another. Because of officials' failure to complete a bypass route around the capital, they are forced through town, where they spew noxious exhaust and create a strange midnight rush hour. Most are poorly maintained, and owners order drivers to cut their diesel with kerosene to save money, adding an extra dose of poison to the exhaust.[21] In winter, the fumes are held in place long into the morning by cold air pressing down over the city, a phenomenon known as temperature inversion.

The rules governing big diesel trucks are on the front pages during my visit, as India's environmental court, the National Green Tribunal, has just barred those more than 10 years old from entering the capital. The intention may be noble, but the implementation is absurd. The new rule, says the tribunal, is to take effect within hours of its judgment being issued, that very night. Not surprisingly, there is chaos and confusion at checkpoints, and soon the truck drivers are threatening to strike. It's easy to see why they're upset: these men, working long hours in a tough job, are likely to pay a steep price to help clean Delhi's air. It's one they can scarcely afford.

* * *

The intersection of environmental regulation and bread-and-butter economic worries is always contentious territory, and after hearing of the strike threat, I want to speak to those at the center of this storm. Once again, Amit knows where to find them. Just outside the city limits, he parks his little white sedan behind a line of hulking trucks and stacks of huge tires, outside a row of doorless shopfronts. One of them is the headquarters of Ajit Jodhpur Roadlines, a small outfit whose trucks carry machinery, food, and other goods between Delhi and Rajasthan, the northwestern desert state. In the stiflingly hot main room, flies buzz everywhere as a dozen young men relax on a raised concrete platform; some are stretched out atop a dirty blanket, others sit cross-legged, waiting for the next job or resting ahead of an all-night drive. Flamboyantly colored pictures of Sikh leaders and Hindu deities are pinned up behind a desk in the corner.

Surinder Singh got in at 3:00 a.m. from Jaipur, 165 miles south, and he'll head back there tonight with a load of iron pipes. He slept in his truck this morning and plans to stop at home for a few hours before his run, he tells me, gripping a folded-up newspaper in one hand, the sleeves of his brown striped shirt rolled up. Singh is a hefty man, with a round face and bushy moustache, white stubble on his cheeks and chin. "It's a tough life, but this is what I do," he says. "There is no rest." He bought his rig 11 years ago and only just finished paying for it. He figures the court's decision has slashed the truck's value by two-thirds. "Nobody will want to buy it." He'll probably stop driving, maybe open a shop or restaurant, but he fears for the future. "If I lose my business, my job, how will I feed my children?" he asks. "How will I educate them?" He knows trucks like his foul the air, but, gesturing toward a garbage fire across the street, says the industry is unfairly scapegoated for a much bigger problem. "Yes, there is pollution" from trucks, he says angrily. "But this is also a way of earning a living."[22]

His bosses are furious about the court order too. "It feels like a betrayal," says Ajit Singh, who wears a bright blue turban and a thick white beard that's braided at the bottom and tucked back into itself. He owns the agency with his brother, Jaswant (they are not related to Surinder). Jaswant Singh says he first heard about the rule change on TV the night it came into force. Like their drivers, the men are wary of government promises to build the road around Delhi. "So many years have passed," Ajit Singh says. "We don't have any faith."

*   *   *

The morass around the bypass project, approved in 2006 then stalled for a decade, points to a larger issue, one that underpins the air pollution crisis and many of India's other big health problems too. It is the country's utterly inadequate infrastructure, a nagging deficit in roads, energy networks, and the like that has long hindered development. Simply put, India lacks the dull but critical systems needed to keep things functioning smoothly. From sewage treatment to fuel distribution, gaps and bottlenecks are everywhere. In the same way that better roads could keep big trucks out of the capital, adequate garbage collection would eliminate the need to burn refuse. More reliable electricity would make diesel generators unnecessary, and better networks for distributing household gas would end the deadly toll of smoky cooking fires. Air pollution isn't the only health hazard created by such failings. Sanitation, for example, is another huge problem: sewage is barely treated, so infection and parasitic disease are rampant.

One such shortfall, invisible but consequential, is in the oil refining sector. Fuel companies, often state owned, are reluctant to spend to upgrade the plants that turn imported crude into gasoline and diesel. That means perennial shortages of high-grade, low-sulfur fuels, making it difficult for officials to tighten pollution rules on cars and trucks. Automakers and refiners point fingers, each saying they would be happy to meet new standards if only the other would do its part. The stalemate has slowed the rollout of stricter rules, and they are often introduced in a strange patchwork that undermines their effectiveness. Tighter requirements are imposed on cars and trucks registered in big cities, but not those in surrounding areas, or on one half of the country but not the other. A registered vehicle can drive anywhere, though, so many of the trucks plowing through Delhi spew much nastier exhaust than would be allowed if they were based there.

The reasons for such deficits are complex, and much debated. For all the attention to its software start-ups and call centers, India's per capita income is just $1,700 a year, a fifth that of China, the country with which it is most often lumped together in discussions of emerging economies and the environment.[23] That poverty is both cause and effect of an infrastructure crisis that makes it difficult to run factories and ship goods. In *Restart: The Last Chance for the Indian Economy*, a book generating chatter while I am in Delhi, journalist Mihir Sharma blames this vicious cycle on several factors, including leaders' reluctance to complete a program of economic liberalization begun in the 1990s, their decades of prioritizing agriculture over industry, and the endless red tape that can make it harder to move goods from one Indian state to another than to transport them internationally.

Along with the infrastructure deficit is governance failure. India has plenty of tough environmental laws, many of them dating to the aftermath of the 1984 Bhopal toxic gas leak, but they are often ignored. So while doing little to stop pollution, the rules create opportunities for corruption among the petty bureaucrats who decide when and whether to enforce them. Local air quality boards are notoriously weak.[24]

It's not entirely officials' fault, says Ambuj Sagar, a policy professor who moved home to Delhi after 20 years in the United States so his research could inform better decision making in India. Many government departments, he says, are badly overburdened and simply don't have the resources to do more. "It's not that they're sitting around twiddling their thumbs," says Sagar, leaning in over an iced coffee when we meet at a café near his office on the campus of the Indian Institute of Technology Delhi. Good government requires the capacity for thinking ahead. "You say, 'We see this emerging as a problem, and let us start planning for it.' Instead of saying, 'OK, we have our backs to the wall, now we have to do something suddenly.'" I can't help thinking of the effective-immediately court order that hit Surinder Singh and his fellow truckers. A more deliberative approach might have softened the blow for them.

The chaotic nature of government action is clearly on view when I sit in on a press conference called by the environment minister. In the face of intense media concern about air quality, he is announcing a slew of new measures, from tighter monitoring of car exhaust to the overhaul of a coal plant just outside the capital. Some of the changes will come into force within weeks—exactly the kind of on-the-fly approach Sagar decries. Journalists are shouting over one another, and the most forceful questioning comes from a woman demanding clarification of an announcement that only cars meeting tight new standards will be allowed registration in the area around Delhi. "Sir!" she shouts, the decision "has caused so much of panic." She presses the minister to explain when the rule will come into effect and whether it will apply to all cars or just new ones, but gets no clear answer.

One sign of leaders' failure to address knotty problems is the courts' central role in environmental policy. The judiciary's achievements are real: it was the Supreme Court that forced the cleanup of Delhi's air in the first years of the new millennium. But judges don't have the technical expertise to make good decisions on complicated environmental questions. And they can't address big-picture issues like transportation planning or use powerful levers such as taxation. What's more, the possibility of their stepping in makes it easier for elected leaders to avoid taking action.

Ritwick Dutta, a litigator who argues before the National Green Tribunal, sees another problem too. He says judges, like the government, are reluctant to target powerful industries, convinced doing so would slow the economic development India so badly needs. So they penalize little guys while ignoring pollution from coal-fired power plants, whose terrible impact on health is well documented. There is, he says, an "obsession with growth, the mistaken notion that this is the only way we can obtain the developmental levels of the West. You find extremely poor standards, extremely poor compliance, and no court, no government is ever willing to shut down a polluting electricity-generating unit."

In India, environmental questions have always been bound up with the pressures of development. With more than 20 percent of the population lacking electricity and some 270 million in desperate poverty, ecological concerns are often subjugated to the need to raise living standards. Many in the Delhi establishment, says Samir Saran, senior fellow at the Observer Research Foundation, a think tank, shrug off the air crisis as a phase through which every growing economy must pass. "And be it the '70s in Germany, or in the U.K., or in New York, or in any other city, you have to get past this bump" on the road to greater wealth. Saran says he could be willing to entertain that notion. "What I'm not willing to buy is a pathetic response frame. Because then neither will it be temporary and neither will there be another side. And that is my worry. To get to the other side, you have to start acting, you have to start putting in place solutions. And I don't see that conversation in our debates."

Coal has been at the center of this argument. In a remarkable shift, India has finally begun, after years of fealty to the fuel, to cancel plans for new coal-fired power plants. And Prime Minister Narendra Modi has gotten behind a big push to expand solar energy, as falling prices have made it unexpectedly affordable. Nonetheless, coal is likely to provide the lion's share of India's electricity for decades to come. And Modi brooks little debate about it, sometimes even shutting down groups that call attention to coal's terrible toll, from the poison it puts into the air to the twisted limbs and mental deficits suffered by those living near mines and power plants.

The even more consequential issue looming over the coal question, of course, is climate change. India is the world's third-largest greenhouse gas producer, and coal is responsible for 40 percent of that footprint. For all his green talk on the international stage, Modi doesn't take kindly to outside pressure to scale back his country's carbon emissions. He's fond of pointing out, rightly, that the West is responsible for the vast majority of the warming gases that have accumulated in the atmosphere since the Industrial Revolution. Now it is India's turn to grow, and he believes those who have built their societies on fossil fuels are in no position to demand that its 1.3 billion people forgo the benefits of development for the sake of the planet. With the average Indian responsible for 1.7 tons of greenhouse gas emissions annually, compared with nearly 17 tons per capita in the United States and more than 6 tons per capita in Europe, it is a hard argument to refute.[25]

It happens that my visit to Delhi coincides with a steep and sudden jump in locals' understanding of the severity of their air problem. Many are especially alarmed by their hometown's place near the top of the WHO's list of the world's most polluted cities. Beijing, famed for its dirty air, is far better off. The ranking helped propel pollution to the top of the news agenda in a city that had long preferred to ignore it. Now, journalists are pushing officials for answers and seem unlikely to let the issue drop.

Leaders reject the comparison to the Chinese capital, but they nonetheless seem to be taking heed. After his chaotic press conference, the then-environment minister, Prakash Javadekar, speaks briefly to me by phone. He blames the Congress Party, swept from power by Modi's 2014 election victory, for neglecting air quality during its decade in office. "We have decided to take real action," he says. "We will win this war against air pollution, and we'll showcase for the world how to do it in the shortest possible time."

Only time will tell whether improvements materialize, but so far, the signs are not good. "If you want to tell me that we mention air pollution every day, yes we do," says Saran, the think tank expert. "But that doesn't mean we are doing anything about it." Only vision and political will can make a real difference, he argues. "What are we doing about the next 1 million vehicles that are coming, or the next 5 million? Because it will happen." The lesson of India's 1990s push on pollution, in his view, is that tough action may raise hackles at first, but is soon accepted. Others wonder whether public awareness, while growing, is yet where it needs to be. Barun Aggarwal says that in his work selling air purifiers, "people still get on the phone, these are educated people who read the newspapers, and they say, 'Oh this is just propaganda by the [filter] manufacturers,'" or "'I've lived here all my life and nothing's happened to me.'"

* * *

One place where alarm is undoubtedly high is inside Delhi's expat bubble. Gen Chase, an American international development consultant who moved to India from a posting in Ethiopia, tells me her family turned down an opportunity to go to Beijing because of its pollution. Although her professional expertise is in public health, Chase says she didn't realize before arriving in India that its air was far worse. "I do research about the place I'm going to before I go there. And this just wasn't in there, it just wasn't anywhere," she says, chatting in an air-conditioned café on the sprawling, manicured campus of the American Embassy School after dropping off her kids one morning. "We came with zero knowledge of the air." That was before the WHO report and the blitz of media attention. "I mean, I wrote letters home to people saying, 'There's this wonderful ethereal mist over everything,'" but she gradually began to wonder what was causing frequent headaches, sore throats, and more among her family and friends.

"For me there was a pivotal moment in that first year, where I was at the school, standing around with a bunch of expat parents, and there was a conversation that everyone felt horrible, felt sick, and there must be some virus going around that affects your throat and your eyes. And I really started asking myself and the people around me, 'Do you think it could it be pollution?'" No one she knew was talking about it, but "I was seeing things in my own kids," who were athletic before coming to Delhi but began scaling back their sports because exercise felt so uncomfortable. Her daughter "used to run off the track and have rosy cheeks and be healthy. Here in Delhi she comes off pale, she doesn't feel good, she feels dizzy."

Chase, who is tall and athletic-looking herself with a chunky wood-bead bracelet and sleeveless linen top, began watching the air quality numbers and keeping her daughter out of track when they spiked. "I was pulling her off all the time," she tells me. Since then, she says, her understanding of the danger has evolved. "Now, knowing that I pulled her off when it was over 500. I mean, are you kidding me?" she laughs. Chase is referring to the Air Quality Index, which incorporates readings of various pollutants; 500 is the upper limit, the zone color-coded maroon for "hazardous." It's not just the horrific days, she realizes now, that cause harm: in Paris, she says, "they shut the city down at 180. I was just looking at the AQI, and it said 190, which is only 'unhealthy.' It's very unhealthy. It's where actually most of the cardiovascular problems are, at the lower end." There are still times, says Chase, when readings are through the roof and "the kids are all at, literally, a seven-school soccer tournament in the air outside, where I've thought, 'We are all insane.'"

It was a powerful article by New York Times correspondent Gardiner Harris, coupled with the WHO ranking, that jolted the expat world—and many beyond it—into an understanding of the severity of Delhi's air problem. "People were really upset about that piece and really interested, like 'Are you kidding? Is that really real?'" Chase recalls. "It breeds panic," she says. "Disbelief and then panic. Then there's denial." Many foreigners—diplomats, development experts, businesspeople—have come to Delhi for their dream jobs, stimulating professional opportunities that can be the pinnacle of a career. "Oh, it's so exciting," Chase says. "The school is wonderful, the work is fascinating, Delhi is beguiling—wonderful, magic, and crazy."

So she's sometimes been tempted to tune out the reality of pollution. But with her husband reaching the end of his three-year commitment at work, they've decided to head home to the United States; not quite an early departure, but a considered decision not to extend. Many others, Chase thinks, have done the same. What's more, colleagues who might have succeeded them are taking Delhi off their lists. "That's huge with people in our world, in development. It's like, 'Yeah, I've been reading the press, I'm not going to bring my kid into that.'" During her time in Delhi, Chase started a website devoted to the issue, and decided to make it her new career focus. More importantly, she says, with India's media on the story, "it's no longer some precious little expat issue, thank God. Because that's not the way you're going to make change."

In a follow-up piece published as he left Delhi, Harris wondered whether it had been ethical to subject his young sons to the city's air. The boys' years in India had put them at risk for lowered lung function—and also, therefore, for disability and early death. One Indian air expert, who moved to Goa, on the coast, to protect his two young children from the capital's pollution, was unequivocal, Harris wrote: "'If you have the option to live elsewhere, you should not raise children in Delhi.'"[26]

Manjali Khosla, whose children hold their noses upon returning to Delhi from vacations in Canada, sees it differently. Her kids also go to the American Embassy School, and she meets me in the campus café after Gen Chase departs. Khosla and her husband moved to Delhi after decades in Hong Kong, Canada, and the United States so the children could experience Indian culture and spend time with their extended family. After a bout with pneumonia, Khosla began exhorting friends and relatives to wear face masks, which are rarely seen in Delhi. "My kids do say, 'Why are we here, why don't we just leave?' And I say, 'If we leave, it doesn't solve the problem. If we leave, your grandparents are here, your cousins are still here. Let's stay and try to fix it. And in the meantime, wear masks.'"

* * *

Most Indians, of course, don't have the choice. On a concrete island in the middle of an eight-lane highway, near a stretch of south Delhi road named for Mahatma Gandhi, I meet Mohammad and Babli Yunus. They're raising their five children on this patch of pavement, surrounded on three sides by the highway and pressed on the fourth against a tall metal fence. Above them, an overpass carries several more lanes of traffic.

The family's possessions hang from the fence's spikes: a plastic bucket, bundles wrapped in patterned cloth, a few blankets, a shirt. Beside the curb, a baby and a toddler sleep on a dirty green mat, a shawl covering their faces. The baby, two or three months old, wriggles a little, and a hand peeks out from under the cloth, a bangle encircling the little wrist. Cars, trucks, and motorcycles roar past just a few feet away, but the children sleep soundly, and before long an older boy, maybe four, wanders over and stretches out beside them. Within a moment, he is asleep too.

As Neha and I speak with the young father, a dozen or so children crowd around, the offspring of neighboring families as well as his own. Mohammad Yunus is a few rungs of the socioeconomic ladder below the auto-rickshaw drivers I met earlier, but his words echo theirs. Unable to afford land or find work farming, he and his family left their village for the capital. He earns what he can as a day laborer, and the children bring in a few rupees collecting discarded plastic bottles to sell.

Families like the Yunuses are everywhere in Delhi, cooking over fires by the sides of roads, camped under plastic tarpaulins at the edges of parks, sleeping by the hundreds outside the famed Jama mosque in the old quarter. Their problems, of course, are legion, and the dirty air they breathe all day and night must be low on the list. No school, no sanitation, no medicine when they get sick; those immediate worries make far greater demands on their attention.

Still, I can't help wondering what terrible mix of toxins poisons the heavy, acrid air here in the middle of the highway. And what invisible havoc it is wreaking on these children, the little ones napping inches from traffic, their bodies made vulnerable by poor diet, lack of shelter, and frequent infections. Babli Yunus wonders too. She has a thick, phlegmy cough that never goes away. "My health is so bad, and my children are also suffering, lying around here. But what can I do?" she says. A daughter, maybe three years old, leans back in her lap, playing with a dirty yellow comb. "This child has a cough and cold. So does this one." Their mother doesn't know whether the latest round of illness was caused by the exhaust or by a soaking rain that left their clothes and blankets wet. "It all goes into our systems," she says. "We end up eating a lot of dust."

The Yunuses have been living in this wretched spot for two or three years, and when Neha asks how much longer they might stay, Mohammad Yunus has no answer. "Where should we go?" he asks me. "You tell us."

Their island in the highway is part of a larger encampment that sprawls alongside the road, home to hundreds of people. All around, as dusk falls, their evening routines unfold. The oldest Yunus son wakes the baby, who looks grumpy and tired in his arms, but doesn't fuss. An older girl in a pink dress pulls sticks from a sack and kindles a fire; when she tosses a couple of plastic bottles into the flames, the smoke turns rancid and heavy. She pulls dough from a big metal bowl and begins shaping it into circular patties, roti bread for tonight's meal.[27]

3

9,416

Living London's Diesel Disaster

Thick and heavy. Steady, reliable. The world's workhorse. Diesel. Its power hidden from sight, deep inside an engine, where air whooshes into a dark chamber. A piston pushes in, compresses the air, confines it to a tiny fraction of the space it filled just a second ago. The pressure jumps, and temperature does too. Hotter and hotter. Hundreds of degrees, a thousand. Now comes the fuel, entering this tight space in a spray. Igniting, instantly. The explosion's power pressing back on the piston, pushing it down. Power unleashed, then harnessed. To turn the wheels. Push the bus up the hill, the truck along the highway. To generate the electricity that gets the hospital through the storm. The piston slides in again, clearing out the dark smoke. As it drifts away, a new puff of air comes into the chamber, and the cycle begins again. Pressure. Explosion. Power. To do the work that needs doing. Dirty work. But essential.

It was a month before my 30th birthday that I moved to London. I'd met Dan a few years earlier, when he was a physics researcher at Princeton and I was living in New York. At a mutual friend's beach house one summer weekend, I found myself following his explanations of string theory with a keener interest than I'd ever imagined having for the subject, and listening to stories of a British childhood that seemed so far removed from my own upbringing in the New Jersey suburbs. Before long, he was visiting most weekends, and I, on the irregular, midweek days off I got from covering New York City news for the Associated Press, rode the train to Princeton. When he was offered a teaching post in England, I persuaded the AP to give me a job there too.

It was December, the worst time to land in London. The steady drizzle felt endless, the sky was gloomy at midday and dark by 4:00 p.m., and despite the street atlas weighing down my bag—this was before smartphones—I was constantly lost on roads whose names seemed to change every few blocks. I missed the logical grid of New York, and much more, too, unsure where I fit in this new home I hadn't quite chosen, a city foreign enough to flummox me but not to provide the thrill of the new I'd found in farther-flung travels through Asia when I was younger. Gradually, though, I began to find my place, to make London my own. I learned how to decode the politely evasive responses of the government officials I interviewed for work, made my peace with the moths that chewed holes in all my favorite sweaters, and set out to build a life with the man I'd followed across an ocean.

But there was one thing that still bothered me. London felt dirty in a way I had never experienced in New York. Its air was thick, and after even a quick dash out of my office for a sandwich, I felt a layer of grit coating my teeth and a faint bitterness in the back of my throat. A longer walk often left me a little light-headed. No one else ever mentioned feeling that way, or seemed at all concerned, so I figured I must be imagining it, and I dismissed the whole thing as a minor annoyance, to be filed in the same category as the inadequacy of the coffee (since improved) and the impenetrable accents I sometimes encountered. Over the years, I occasionally heard Londoners joke that breathing in the city could take years off your life, but no one seemed to take such notions seriously, and of course the famous pea-soup fogs were the stuff of history.

It wasn't until much later, while researching an article about the soon-to-begin 2012 Olympics, that I had occasion to question that complacency. By that time, we'd had our daughter, so I had Anna's health in mind, as well as Dan's and my own, when I went online to read the science of pollution's effects. It took just a few minutes to shatter my comfortable illusion that London's dirty air was no more than a nuisance. Although few seemed to realize it, this city—indeed, the entire country, along with much of the rest of Europe—was in the midst of a full-blown public health crisis.

* * *

To grab ahold of the story, start with a number: 9,416. Astonishing in its specificity, it's the culmination of an upside-down murder mystery in which the perpetrator's identity was clear from the start, but no one knew who the victims were. It's not their names, though, that are at the center of this tale. What the detectives were searching for was a number, and 9,416 was the one they found. That was the answer—one answer, anyway—to the question of how many victims there were. How many Londoners die because of air pollution each year.

Its precision is misleading, because 9,416 is not a body count, a tally of individual diagnoses. But it is grounded, nonetheless, in scientific rigor, the product of a careful application of what is known about pollution's power to the particulars of this city's air and its people. It's a maximum, the high end of a range of possibility. And it could evolve further; even by the time you read this, 9,416 may have been replaced by a new number, an updated or reconsidered estimate of pollution's casualties.

In any case, a death toll is not the best way to express the damage dirty air wreaks. But even those cloistered in the lab can see the power of a number like 9,416. The power to make headlines, certainly, and to put pressure on politicians. But more than that, the power to haunt the imaginations of Londoners like me, to grab us—the people who breathe this stuff—by the shoulders and shake us, to make us fear we or someone we love may become one of the 9,416.

* * *

Long before 9,416 existed, Simon Birkett knew what it would mean to have a number like that. Birkett's not a scientist, but since leaving his banking job in 2009 to throw himself into this work full-time, he's acquired a deep expertise on London's air. Not just the details of NOx and PM2.5 and the other contaminants, but the intricacies of the rules that govern what pollution is allowed, and the inner workings of the bureaucracies tasked with enforcing those rules.

Birkett is single-minded about this issue. Obsessive, really. Working mostly on his own on a campaign he named Clean Air in London, he's spent years writing letters to government ministers, members of Parliament, and EU officials, bombarding them with Freedom of Information requests and talking to anyone who will listen about the invisible health crisis gripping his city. We'd spoken on the phone a few times, exchanged many emails and Twitter messages, but I hadn't met Simon Birkett in person until he answered the door of his little yellow cottage one early spring afternoon and invited me inside. We settle in to talk in an upstairs sitting room that overlooks a quiet lane, a tranquil view just a block from a busy urban thoroughfare, a few minutes' walk from Harrods department store.

Birkett is bald on top, with gray hair trimmed close at the sides and heavy-framed black glasses. As he speaks, he repeatedly picks up his phone to send me reports backing up something he's said, or just footnotes himself verbally: "This is all on the website." Indeed, his site is an exhaustive compendium of facts, figures, and official reports. It was in one such paper, from 2005, that Mayor Ken Livingstone mentioned the number that drew him into the statistical chase. Back then, it was 1,031, and Birkett wanted to know how that estimate of London's pollution death count had been derived. He filed a Freedom of Information request, but answers were elusive. Everyone "kept fobbing me off and fobbing me off."

Later on, in 2009, he did his own calculations, drawing on a new EU finding that, for every million Britons over 30, pollution caused 650 premature deaths a year.[1] That gave him a London death estimate that nearly tripled the existing figure. Birkett's number—and he'd assumed, very conservatively, that the capital's diesel-fogged air was no worse than the national average—was 2,905. So he composed a 13-page letter to Britain's health secretary (the last 7 pages were actually a long list of everyone he'd copied in, plus appendices full of links to relevant reports), laying out his calculations and demanding government come clean about the impact of London's pollution.[2]

Eventually, he was invited to the Department for Environment, Food and Rural Affairs, where he finally got an answer to his original question. "They had a laptop, and they turned it 'round and they said,

'That's where the 1,031 comes from.' It was just a cell in a spreadsheet." Birkett understood what he was looking at, and he immediately spotted a problem. In this field, deaths are generally estimated by using a percentage, known as a coefficient, indicating the increase expected to result from a given rise in pollution levels. But he could see they'd used the wrong coefficient, one that told only part of the story. "That's short-term exposure," he remembers saying, a number that captured only deaths occurring right after a victim breathed pollution. The roomful of government experts, in Simon Birkett's telling, were befuddled. But there was one person "who knew exactly what I was talking about." That was Heather Walton, then a toxicologist at the Health Protection Agency, who, as Birkett recalls it, was conferencing in by phone. Walton told me she didn't remember the conversation, but later on, she would play a central role in the story too.

In the meantime, Birkett got back to work. An American expert helped him do some new calculations, using a long-term coefficient. "That's when it all started to come out," he says. His new estimate was a range, between 3,459 and 7,900 Londoners dead annually from particle pollution. He put his figures into another scathing letter, this one to the House of Commons' Environmental Audit Committee. The government's failure to warn Britons about pollution's effects, he wrote, "may represent one of the biggest public health failings or 'cover-ups' in modern history."[3]

When Frank Kelly of King's College London, one of the country's top pollution scientists, was scheduled to testify before the committee, Birkett suggested lawmakers ask his view of the new number. "There are all these MPs, half asleep, and one of them says, 'Well, Clean Air in London says this. What do you think, Professor Kelly?' And of course, they were all expecting him to say, 'This is just Birkett being hysterical.'" Instead, Kelly agreed the estimates seemed reasonable.[4] "Once you'd had somebody as respected as Frank saying, 'Those numbers are about right,'" Birkett tells me, "that blew the lid off."

Boris Johnson had by then succeeded Livingstone as mayor, and he published a new study, with an updated number for London: 4,267.[5] Birkett's calculations hadn't been far off. Meanwhile, Heather Walton, the voice on the other end of the phone when he'd glimpsed the government spreadsheet, had moved to King's College, where she'd joined Frank Kelly's group and was running a study that would revisit the London number yet again, this time adding the toll from the capital's worryingly high nitrogen dioxide pollution to estimates that had until then considered only particles. By the time they finished, in 2015, she and her colleagues had come up with the most alarming estimate yet: 9,416.

* * *

One of the many shocking things about 9,416 is that it represents nearly 20 percent of London's annual deaths.[6] A fifth of those who die in this city, in other words, may have had their end hastened by dirty air. One day, Heather Walton was waiting for a train, and she overheard fellow passengers discussing that very piece of news. "That can't be right," one said to the other. "I would have noticed it." Walton couldn't help interjecting. "Actually, I wrote that report," she told them, and proceeded to explain some of the caveats around the number. She worries news reporting that neglects to do so—and it's impossible to capture the complexities of statistical techniques in a headline—just makes people distrustful of science. Yet, like many in her field, Walton is caught between the carefully couched detail of scientific papers most people will never read and the need to present findings in a way the wider public can understand.

The thing that made her work groundbreaking, that pushed her number so much higher than its predecessors, was also what added to its uncertainty. London has a serious nitrogen dioxide problem, because the pollutant pours from the diesel cars Britain has foolishly encouraged its people to buy. While less dangerous than particulates, NOx is present here at far higher levels. Previous death estimates hadn't accounted for it, because research on its effects wasn't as advanced. One important, and unanswered, question is how much the damage wrought by the two pollutants overlaps. In other words, should the deaths caused by nitrogen dioxide simply be added to those resulting from particles, or are they the same deaths? Because no one's quite sure, Walton's estimate is a range rather than a single figure. At one extreme, "they overlap completely," and at the other, "they don't overlap at all. And you know neither of those are likely to be correct." After she built in some overlap, the number at the top of her range was 9,416. So, preceding it with phrases like "could be as high as" that most readers breeze right past, that's the one we journalists grab hold of.

It's not that 9,416 is wrong, or pulled from thin air. On the contrary, as I sit in Walton's office near Waterloo Station, I'm more impressed than ever with the painstaking way in which she and her colleagues arrived at it. She explains how they broke London down into squares just 20 meters by 20 meters and created estimates of the pollution levels in every single one. They matched those against population records to figure out how many people were breathing each square's air. Then they used a rigorous estimate of pollution's effects, from a 16-year study of 1.2 million Americans: a 6 percent jump in deaths for every 10-microgram-per-cubic-meter increase in fine particulates.[7]

Another haunting figure from Heather Walton's study is the amount of life a typical Londoner loses because of air pollution: about 9 months from particulates and as much as 17 months from NOx (although she warns they shouldn't be added together). But those are averages, and it's hard to know how the loss is really distributed. Do a relatively small number of people die decades earlier than they might have, or do many of us lose a shorter time? For now, the experts can't say.

While these kinds of numbers have long eaten at me, I realize yet again, in talking to Walton, just how hard it is for us nonscientists to know what to do with them. Sometimes I can talk myself into believing the time pollution may take from me isn't much at all compared with what I'll lose for not being a more diligent exerciser, or showing a little willpower when the sweets in the cupboard call my name. Hopefully it's a long way off anyhow, and maybe we'll live someplace else by then. I've sometimes caught myself calculating that if one in five Londoners dies from air pollution, perhaps the odds will favor my little family, and we'll escape its effects altogether. Then all my worrying will have been for nothing. On the other hand, maybe one or even two of us three will fall into the unlucky 20 percent.

Needless to say, while such rationalizing may be natural, that's not how it works. Numbers like the ones from Heather Walton's study are intended to express risks to an entire population—a city, a country, millions of people—not an individual. They convey something urgent and vital, giving us a sense of a problem's scale so we can compare it with other dangers and decide whether to do something about it. But sometimes it's hard to take them the way they're intended, to resist twisting the frightening figures into something that speaks to us more personally.

David Spiegelhalter is a professor of the public understanding of risk at Cambridge University, so he knows that better than most. Estimating pollution deaths is tricky, he explains when we speak on the phone, and helping people know what to make of the numbers is even harder. He thinks Walton shouldn't have published 9,416, because rounding it to "about 9,000" or "nearly 10,000," would have given a clearer sense of its roughness. That's how a committee reporting nationally on the problem in 2010 handled it, estimating the effect of particle pollution on Britons as equivalent to nearly 29,000 deaths a year.[8] When the royal colleges of physicians and pediatrics later incorporated the effects of NOx, the annual toll jumped to 40,000 (and the economic impact to £22 billion).[9] In fact, even that level of precision is, in Spiegelhalter's view, not quite right. When he sees 40,000, "I just think 'tens of thousands.' It's got four noughts on it," he says, using a British word for zeros. "America's got five noughts on it, and China's got six noughts on it."

That notion, that these death estimates are best taken as broad indicators of the magnitude of a particular locale's pollution problem, is something I've intuited on my own as I've encountered them in my reporting. Even for a single country, they can vary widely, depending on the pollutants they include and the methodology they use. These numbers are everywhere: more than a million and a half annual air pollution deaths each for China and India.[10] Approaching a half million in Europe.[11] Upward of a hundred thousand in America.[12]

None are arrived at by counting individual cases; like Walton's, they're all derived through complex statistical modeling. Even if you tried, David Spiegelhalter says, it would be impossible to compile a body-by-body tabulation, since pollution—unlike, say, a heart attack or stroke—is not a cause of death in the medical sense. It's more akin to smoking, obesity, or inactivity, all risk factors that can hasten a death or make it more likely, either alone or as one of several contributing factors.

That's not to say there's no point trying to sketch the outlines of its impact, Spiegelhalter is quick to add. Doing so allows us to see how it stacks up against other risks. One of the notable things about the U.K.-wide pollution estimates is that they put dirty air in the very top tier of this country's killers, second only to smoking. After years of thinking of tobacco, alcohol, being overweight, and inactivity as the big four health threats, "to see air pollution plonked into the middle of them is quite a shock," Spiegelhalter says. While he thinks precise rankings are unreliable, "these are all four-zeros problems," taking tens of thousands of British lives a year.

The sense of scale is important because it helps us set priorities. Some threats, Spiegelhalter says—"exposures that haven't got any noughts at all"—get attention far out of proportion to their impact. He mentions nuclear waste: "You can't find anyone that's been harmed by it." The impossibility of enumerating dirty air's effects perfectly doesn't mean it's not a serious threat. "What it means is it's extremely difficult to say exactly how big the problem is. But it's a big problem."

That, in his view, is the point. Scientific uncertainty has become a politically loaded concept, easily misunderstood and seized upon as weakness by those who don't like scientists' conclusions. This dynamic is most highly charged, of course, around climate change. Those who refuse to acknowledge the overwhelming consensus pounce whenever they hear the word "uncertain." But being unsure whether this century's sea level rise will be three feet or six feet does not mean the seas aren't rising, or the climate's not changing. The same goes for air pollution. "Being uncertain doesn't mean you don't know anything," Spiegelhalter says. "It just means you don't know everything."

* * *

As a young man in the 1880s, Rudolf Diesel worked as a refrigeration engineer. But his real passion was outside the office: a quest to build an engine that could run more efficiently than the steam- and gasoline-powered ones then competing to take the place of horses. Instead of compressing fuel and air together, and then igniting it with a spark plug, his engine would compress only the air, until it was hot enough to ignite the fuel on its own. It didn't always work, and some of his earliest customers demanded their money back. But he kept tinkering, and soon his engine, reliable and slow to wear out, had become a ubiquitous workhorse.

Diesel was German, and it was his country's automakers who would, in the years to come, embrace his invention most enthusiastically. But there were dangers he hadn't anticipated. The engine's intense heat delivered the efficiency he'd dreamed of, but it also prompted nitrogen in the air to react with oxygen, creating gases known as nitrogen oxides—NOx for short. The grouping includes both nitric oxide and its more menacing sibling, nitrogen dioxide, which inflames the airways and combines with other emissions to create still more pollutants, mainly particles and ozone.

Diesel believed his invention might run best on vegetable oil, and he used peanut oil in a demonstration at the 1900 World's Fair in Paris. So it was one of history's small injustices that his name was given to a fuel derived from petroleum, a dense one, rich with energy. Unfortunately for his legacy, it was rife with chemical impurities, too, so burning it would create clouds of toxic particles as well as NOx.

When Diesel vanished en route from Belgium to London in 1913, the details of his disappearance filled front pages: his coat left neatly on ship's deck in the English Channel, pajamas laid out on the bed in his cabin. There was speculation the oil industry might have done him in because of his advocacy for alternative fuels. Others thought, with war coming, Germany wanted to stop him from selling his technology to the British. More likely, it was suicide; despite the great wealth his engine had brought, he seemed unable to get out from under his debts.

A century later, his mysterious demise long forgotten, Rudolf Diesel's name is at the center of another story, one with consequences for many more lives than just his own. Today, diesel is the poison choking Europe. It's well suited to heavy work, so elsewhere in the world it powers mostly buses and trucks. But in recent decades, the continent Diesel called home has embraced and encouraged it for cars too. In Britain, there were 1.6 million of them on the roads in 1994; 20 years later, that number had jumped to almost 11 million.[13]

A diesel engine's distinctive chug-chugging, I've learned from experience, is reliable warning you're about to get a face full of exhaust. Their ubiquity—in London's red double-decker buses, big black cabs, to-your-doorstep delivery vans, and hundreds of thousands of cars—is the cause of what's become a depressingly reliable welcome to every new year. Before the end of January—sometimes less than a week in—local news reports that the most traffic-clogged roads have already exhausted their annual allowance of nitrogen dioxide violations. Some will breach the limit well over a thousand times by December's end.

The story is much bigger than one city. This health threat spans a continent. Birmingham and Leeds, Paris and Rome, Budapest and Prague: All are crowded with diesels, and sputtering on the fumes. Stuttgart, tagged with an unhappy reputation as Germany's Beijing. Madrid, where pollutants bake in the sun and an environmental group gave authorities an "F" for failing to confront the problem. Glasgow and Vienna, both with particulate levels higher than London's. The French Alps, where wood smoke compounds the diesel taint and craggy mountains hold a blanket of pollution in place.

Indeed, it's not only big cities that are affected, as I was reminded on a trip to soggy, beautiful Yorkshire for an old friend's birthday celebration. Both the train that took us north and the taxi we climbed into for a drive over winding country roads gave off the same noisy rumble, the same unpleasant fumes. Even in a tiny, picture-perfect village where we stopped once for tea and cake on a visit to Exmoor, in southwestern England, I could taste the exhaust every time a car meandered by. It's often said that air pollution is invisible, insidious, wreaking harm unbeknownst to its victims. But this is also true: Diesel smells. Its smoke is thick and chalky, giving weight to the air.

Recently, I've started to notice the layer of gray haze when I descend into Heathrow. Glimpsing this man-made poison from above is clarifying, the dangers of a toxic cloud engulfing a place where millions of people live painfully obvious. Back on the ground, though, that sharp focus blurs, the threat fading into the background.

Air pollution has a long history here, of course. The Great Smog of 1952 ranks among history's worst environmental disasters; it killed an estimated 12,000 Londoners. They crowded hospital wards and died in their homes, some during the five days when dark fog sat so heavily it was hard to see, others in the weeks that followed. It was the culmination of a century and a half of "pea soupers" that had become part of London lore: thick yellow fogs that, while they wrecked health and lives, were romanticized on canvas by Monet and Whistler and immortalized on the page by Dickens, Conrad, and Arthur Conan Doyle.[14]

Back then, coal was the cause of the city's suffering, the black smoke pouring from every chimney, and from the smokestacks of power plants. The 1956 Clean Air Act, passed in response to the Great Smog, put new restrictions on urban coal burning and was the first step toward ending a terrible era.

Today, the challenge is different. London's pollution is, of course, far less dire than Delhi's, Beijing's, or Krakow's. But a more fitting comparison, a global peer, is New York, a city similarly plagued by traffic—but, with little of it diesel, air much cleaner than London's. Despite the huge SUVs crowding U.S. roads, Britons are more than twice as likely as Americans to die from the effects of dirty air.[15]

*　*　*

Frank Kelly understands that troubling reality better than most. As head of the Environmental Research Group at King's College, he is one of the foremost experts on London's pollution. But he didn't start out studying air. Early on, his focus was lung damage in premature babies. Later, he began looking at other breathing ailments, too, from asthma to cystic fibrosis. Many, he saw, were caused or worsened by damage the body inflicted on itself through an overzealous, or misdirected, immune response.

It was after he moved to London in the early 1990s that he became interested in pollution. When he began peering inside the human body, he tells me, dirty air's effects were plain to see. Londoners' lungs were too compromised to study, so Kelly, a solidly built man with thick salt-and-pepper hair and a Belfast lilt in his voice, did his work in northern Sweden. Volunteers spent two hours in chambers filled with diesel fumes, at levels similar to those on Oxford Street, the shopping strip whose crush of big red buses makes it one of London's worst pollution hotspots.

When the subjects emerged, researchers used a tool called a bronchoscope to take fluid and tissue samples from their lungs. The lining of the lungs and airways, Kelly tells me, "is a war zone," the staging ground for the body's relentless fight to protect itself from the outside world. But when the invader was man-made pollution rather than a natural threat, he found, the system turned against itself.

Volunteers' bodies were responding to the contaminants as they would an invading bacteria or virus. Their immune systems kicked into gear, and defensive cells rushed toward the battle. "What they come across is a little black soot particle covered in lots of chemicals, which they've got no way of dealing with." So they attack the lung itself, doing a little more damage each time. "If you live in a city, and you breathe in that level of pollution" year after year, the damage accumulates. Eventually, the tissue can stop regenerating, and the lung struggles to do its job.

Kelly's team was also out on the streets, measuring pollution levels, and the more they learned, the clearer the picture became. Along its busiest roads, London "is a gas chamber," he tells me. "We've walked into this sort of nightmare." Not only does diesel produce more NOx and particles than gasoline, those particles are coated with a nastier brew of chemicals, "more varied and more toxic." In fact, Kelly says, because today's pollution particles are so small, penetrating the body so deeply, London's air may be more dangerous now than in the 1950s.

It's not only the stuff emitted on London's own streets that's fouling its air. With monitoring stations across southern England, Kelly's group tracks pollution drifting in from elsewhere in Europe. If at 6:00 a.m. the levels rise near the Channel, they'll spike in London by 11:00 a.m. Using wind and weather data, the scientists trace the trajectory back to France, the Netherlands, even as far as Poland. "If the weather's of the right nature, half the continent gets pollution from Poland" and its relentless coal burning, he tells me.

So Frank Kelly knows London can't fix this alone. It's up to Britain's leaders to end the country's love affair with diesel, and put their muscle behind bigger changes too. The same goes for Brussels, because as long as French traffic exhaust, Polish coal smoke, and ammonia wafting from Dutch farms are blowing across the water, neither Britons nor their European neighbors will breathe clean air.

Another source, one in which I feel especially complicit, is flying. One study found pollution from aircraft adds 10,000 deaths to the worldwide toll each year—not much compared with other contributors, but certainly significant.[16] The bigger concern about aviation is its effect on the climate. Flying is a major, and rapidly growing, source of greenhouse gases, and it's undoubtedly my own environmental Achilles' heel. One statistic I came across hit particularly close to home: Every round trip between London and New York causes about 32 square feet of Arctic Sea ice to melt.[17] I can't even count the number of times I've made that journey since I moved to Britain. And of course I've taken even more long-haul flights for this book, adding to the burden on the climate while trying to shine a light on it.

Change won't be easy. At home, it must start with better cars, buses, and trucks on London's roads, but Kelly says that won't be enough. "The big elephant in the room" is that there must be fewer cars all told, maybe half as many as today. That means a fundamental rethinking of how Londoners get around—big investment in the creaking public transportation network, tougher restrictions on driving. "I'd reintroduce an electric tram system across the city. I'd clean up the Underground, improve the frequency of trains," pedestrianize large areas. He invokes the spectacular renovation of King's Cross Station and revitalization of the once-seedy neighborhood surrounding it. If Kelly was in charge, he'd "make sure everything looks like a King's Cross," clean and attractive. There and at stations around the city, commuters would be able to walk outside and "either jump on an electric tram or get on their bike."

It's an appealing vision, one not beyond our reach. Britain, in the years I've been here, has shown itself more than able to do big things—in addition to the new King's Cross, there's Crossrail, the Olympic Park project, the Jubilee Line extension, and more. Small shifts are already visible: Cycling, for example, has been steadily gaining popularity. The bigger changes will take commitment, and money. But they're eminently doable. Frank Kelly's right about one thing, though. It took a long time to create today's mess, and with all the diesel buses, trucks, vans, and cars on Britain's streets, "it's going to take much longer than most people appreciate to get it sorted out."

* * *

I'm a little unsure how to address Sir David King when I email asking for an interview. King was the British government's chief scientific adviser in the Tony Blair years, and I'm eager for his memories of the decision that cast this country's lot with diesel. He's a chemist who taught at Cambridge for years, so Professor should be the obvious choice, but when I'm momentarily unable to find his university web page to confirm that title, I decide against it. And David seems overly familiar. Sir David is too formal—silly, really—but somehow, it's where I land. When he replies the next day, from Singapore, happy to meet as soon as he's home, he signs off as Dave. I can't downshift that abruptly, but Sir David now feels cringe-worthy. So I forgo any address in my reply, opening simply with an expression of thanks.

A bit over a week later, I'm sitting in King's front room in Cambridge, an hour by train from my home. Next to a desk stacked with books and papers, he leans into the side of his chair, arms folded. His sleeves are rolled up, collar open; when he shifts position, one colorfully clad foot pops into view, a blue striped sock with a splash of bright red over the toes. King's name was in the news when I first arrived in Britain. He was the Blair government's public voice during the foot-and-mouth disease outbreak that forced the culling of millions of cows, sheep, and pigs in 2001—and got me sent to northern England to interview the farmers and hotel owners whose livelihoods were ravaged. King tells me it was his advice in that crisis that won him Blair's trust, and meant he had the prime minister's ear when new dilemmas arose.

Amid the daily pressures of governing, King wanted to push officials to think seriously about longer-term risks, and climate change was at the top of his list. He produced a report on flooding—prescient, it turned out—warning that new defenses would be needed as sea levels rose. He was also determined the nation should do its part to lessen the need for such protective measures by confronting the underlying problem. Britain, he believed, should be a leader on climate change.

The prime minister seemed to agree. Even with the wound of the September 11th attacks in America still fresh, Blair called climate change a greater threat than terrorism. He said Britain would cut carbon emissions 60 percent by midcentury (a goal later increased to 80 percent). To do that, King knew, every sector of the economy would have to contribute. Ultimately, the country would have to kick fossil fuels entirely, and that would mean, among other things, moving to radically cleaner cars, powered by electricity or hydrogen.

But the cars of the future weren't ready just yet. The cars of the present would have to suffice a while longer, and shrinking their carbon footprint by even a little bit would help. Diesel was an intriguing option, offering an important advantage over gasoline: better mileage. Driving the same distance on less fuel means less carbon dioxide, perhaps a 17 percent savings.[18] Shortly before King joined Blair's team, government had made a fateful decision to change the way cars were taxed to encourage buyers to choose more efficient models, a change that would effectively incentivize diesel.

As that new tax system kicked in, and officials tinkered with it, King knew his voice might be strong enough to prompt a change of course. And he wasn't yet ready to give his imprimatur to the changes taking root. Diesel might mean less carbon, but he knew it was dirtier than gasoline when it came to the pollutants that threaten not the planet, but the human body. The World Health Organization wouldn't bump diesel up from probable to definite carcinogen until 2012, but there was already plenty of evidence of its dangers.

This was an area where the chief scientific adviser had personal expertise. He'd spent much of his career working with catalytic converters, the most effective pollution control devices in existence, and he knew how dramatically they'd cleaned up gasoline cars. He knew, too, that they weren't nearly as effective on diesel engines, whose sooty exhaust clogs up the devices' tiny channels and whose huge NOx output overwhelms them.

But work was under way on a new technology that looked like it might, finally, clean up diesel. If that promise was fulfilled, Britain could reap this affordable fuel's climate benefit without inflicting a health hazard on its people. So King went to Royston, near Cambridge, where a company called Johnson Matthey had come up with an approach that seemed to do the job. He visited again and again, questioned researchers, and watched as they put one car after another on their lab's test beds—rollers that function like a treadmill—to measure emissions. "It was amazingly successful," he recalls. King had been in California when it mandated catalytic converters in the 1970s, and he saw the rapid clearing of the skies that followed, a change later repeated around the world. "It's one of the modern miracles, getting rid of all that pollution," he tells me. Decades later, a continent and ocean away, "I thought we were going through the same process with diesel."

Diesel, he concluded, was now ready to meet the tougher standards Europe was preparing to impose. Unlike in the United States, where the amount of pollution a vehicle is allowed to emit is the same no matter what kind of fuel it uses, European rules had long given diesels an easier bar to clear than gasoline cars. But tighter limits were coming, and diesel, at last, would have to be nearly as clean as gas. To King, it seemed clear that was finally possible. Technologically, at least, "I thought it was done."

So he was ready to endorse the new incentive. King understood tax breaks for fuel-efficient cars would push the number of diesels on Britain's roads inexorably upward. "We felt that we were going to lead the world in terms of the use of diesel," a clean, modern version. No one saw diesel as the final fix when it came to climate, but it could make a quick down payment on needed carbon reductions. And with the health worries resolved, there was no reason not to move forward.

Across Europe, governments were making similar decisions, seduced by the chimera of a cheap, easy way to cut carbon, and encouraged by powerful automakers deeply invested in diesel. They came to those choices from different histories. Germany, for example, originally cut taxes on diesel fuel to boost its manufacturing sector's competitiveness by making trucking cheaper, then embraced the climate claims to keep the incentive. The impact was dramatic. In 2000, 14 percent of cars bought in Britain had been diesel; by 2012, just over half were.[19] Elsewhere, the embrace was even warmer: In France, 72 percent of new cars purchased were diesel; in Spain, 70 percent; in Italy, 55 percent.[20] In the United States, the land of cheap fuel, where drivers rarely fret over mileage—and NOx rules have been much tighter—less than 2 percent of cars are diesel.[21]

It was around 2011 that King, long since departed from his job advising Tony Blair, first realized something was wrong. Despite the tightening of Europe's pollution rules, air in cities like London wasn't getting any cleaner. At first, he wondered if the new catalytic converters were wearing out faster than anticipated. The real explanation, it turned out, was less innocent. The impressive results King had seen in the lab weren't being matched on the road. As the world now knows, that happened not by accident, but on purpose. Car manufacturers had designed vehicles to run clean while they were being tested and spew filth the rest of the time. So diesel cars that had appeared to satisfy pollution rules were producing seven times the legal nitrogen dioxide limit.

The gaping discrepancy between pollution in the lab and on the roads is at the heart of a staggering tale of corporate malfeasance, one we'll come to in a later chapter. For now, suffice it to say that the unforeseen consequences of the decision to entice British drivers toward diesel would be vastly compounded by the cheating of car manufacturers—and the enforcement failure that let them get away with it. Nowhere has that cheating had deadlier consequences than Europe, which had chosen to depend on a fuel that, it turns out, is still waiting for the revolutionary cleanup David King thought he'd caught sight of all those years ago.

* * *

Even more than King, Damian McBride had a front-row seat to the decisions that set Britain on the road to its diesel disaster. But he remembers things a little differently. Later on, McBride would become infamous as a political hit man, a practitioner of the roughest form of the game, forced off the field when his emails plotting to fabricate sex scandals to smear opponents of his boss, Prime Minister Gordon Brown, became public in 2009. But long before he made it to 10 Downing Street, McBride was a junior civil servant, immersed in the nitty-gritty of tax policy, when Brown held Britain's purse strings as Chancellor of the Exchequer. It is the Treasury that controls taxation, including the car tax at the center of the diesel mess. And Damian McBride says more than just worries about climate change went into the fateful decision to restructure it.

I knew McBride's name, but hadn't been aware of his pre–Downing Street specialization until a politically connected friend mentioned it. McBride turns out to live in the same corner of London I do, and when I ask to meet, he suggests a pub not far from my house. It's hot outside, and the windows have been thrown open; the noise of buses rumbling by, and their thick fumes, fill the place, and we find a table in the corner. He's sweet and friendly, regaling me with a story from his Westminster days as we wait for our drinks. McBride is still young, but he's recuperating from surgery and seems almost frail, his walk a halting shuffle. He hardly seems to fit the role of political brawler. Indeed, he's expressed remorse for his misdeeds, seeming most disappointed in himself for failing Gordon Brown, whom he respected deeply.

Blair and Brown swept into power in 1997, after 18 years of Conservative Party domination, and they were determined not to repeat Labour's old mistakes. They were committed, too, to rebuilding Britain's public services, its schools and hospitals. But they'd promised not to raise income taxes, and the money had to come from somewhere. McBride says fuel duty—the tax levied at the pump on both gasoline and diesel—was central to those plans. Brown hiked it in each of Labour's first three years in power, and drivers didn't take much notice. Fuel duty is regarded as something of a stealth tax, since the cost of filling a tank fluctuates anyway, and Brown's Conservative predecessors had also found it a useful piggy bank.

No one wanted to be accused of punishing drivers, so as they raised fuel duty, officials put McBride to work on cutting the annual tax car owners paid on their vehicles. They built a new system around the goal of promoting fuel efficiency, creating tax brackets, or bands, based on how much carbon dioxide a car emitted per kilometer driven. Those burning less fuel emitted less carbon, so incurred less tax. Ministers boasted they were helping both drivers and the climate, but McBride calls that explanation "rubbish," nothing more than a political cover story.

It wasn't that Tony Blair's government wasn't serious about climate, he says. It was. But its policies, in McBride's telling, fell into two categories: those that really mattered in cutting greenhouse gas emissions, and those designed mainly to look good. Fuel duty fell into the first category; it would limit driving by making it more expensive. Car tax, on the other hand, was mainly window dressing. The Treasury, McBride tells me, didn't expect it to have a major effect on carbon footprints, but thought it would make the other policies go down easier, a useful defense against accusations that Labour was waging war on motorists.

McBride says his team hadn't exactly set out to incentivize diesel, but they realized the changes would have that effect. They knew the health risks, so added a small surcharge for diesels, to weaken the inducement to buy them. But "we were kind of having it both ways." There had been criticism early on that the new system would give breaks only for expensive hybrids. "So we would almost fall off our chairs celebrating that there was a Ford Fiesta diesel that you could buy at one of our lower rates." In fact, just about the only affordable cars in the low-tax band were diesel.

In pounds and pence, the tax break didn't actually amount to much, but it had a power greater than its monetary value. "Mood music,"

McBride calls it. Auto companies began touting the savings, and ministers boasted about it on TV. "Car dealers couldn't pay for that kind of publicity."

Another consequential change soon followed. By 2000, oil prices were spiking, and the fuel duty that once seemed invisible had become a focus of public anger. Enraged truckers blockaded refineries, disrupting supplies, so gas stations had to ration fuel. The effects spilled across the economy: supermarket shelves ran bare, hospitals canceled operations, ambulance crews scaled back service.

On top of the dangers it posed to the country, this crisis threatened the essence of the political brand Blair and Brown had built. It looked, McBride recalls, "like Labour had lost control." Powerful newspapers demanded the fuel tax be cut. So did the Conservatives, who pulled ahead in polls for the first time in eight years. That "played into all the fears that Tony Blair and Gordon Brown had, that they were following the footsteps of previous Labour governments," unable to wield power effectively, and poised to lose it.

The Treasury decided to cut fuel duty, but couldn't seem to be capitulating to the truckers or tabloids. Nor did it want its green credentials questioned. So, McBride tells me over the strange midday dance music blaring in the pub where we sit, he was ordered to come up with an environmental rationale. He drafted a memo outlining a reduction in gasoline tax. It would apply only to a low-sulfur variety, so could be explained, fairly, as a way to shift to that less polluting gas. But to satisfy critics, diesel would get the break too. "That," he says, "was the nakedly political bit."

Brown declined to speak to me, but his statements at the time contradict McBride's story. However the fuel duty decision was arrived at, there's no denying it amounted to another incentive for diesel. When government taxes the two fuels equally, diesel's mileage advantage effectively makes it cheaper than gas. Britain was on its way to today's air quality crisis.

The tragic postscript to King's and McBride's stories is the dawning realization that the promised climate benefit never materialized. It turns out diesel cars' greater weight often wipes out their efficiency gains.[22] And while gasoline cars grew more efficient, carmakers started building bigger, more powerful diesels, so their carbon emissions rose. What's more, diesel's sooty particles—they're known as "black carbon"—trap heat in the atmosphere and are now thought to play an important role in global warming.[23] So for all the damage it's wrought, the diesel revolution failed to deliver on the promise that, while it wouldn't have erased the suffering, might at least have provided a silver lining.

Perhaps it was the humiliating end to Damian McBride's Downing Street career that changed his perspective. Looking back, what seems clear to him now is that while making choices that would affect millions of people, over decades, neither he nor those around him stopped to take a longer-term view. It's especially upsetting, he says, to recall that he and his colleagues were aware of diesel's dangers. Concerned enough to add a nominal premium into their tax overhaul, but not enough to rethink the system they were creating.

More than any one decision, it's that short-sightedness he regrets now: "It's a horrible thing to say, but it almost didn't seem relevant to ask what the long-term consequences were. Didn't seem like that's what my job was." In his view—and I agree—that flaw wasn't unique to one party, or one time. Nor even one country. It's "bound up in the nature of our politics. You're operating one budget to the next, one election to the next," he says. "The mistake is the way we were doing the job."

*   *   *

That lack of long-term perspective was just what David King had hoped to remedy when he agreed to advise Tony Blair. Damian McBride's recollections, of course, are very much at odds with his. That doesn't mean someone's lying. Different parts of government likely saw things differently, were driven by divergent motivations. At the Treasury, McBride was in close proximity to those responsible for the tax decisions, and worked to carry them out. King was much higher-ranking, but not directly involved with tax, and consulted only after the fact.

So who, ultimately, is responsible for Britain's diesel mess? Gordon Brown, the chancellor with the last word on all taxes? Blair, at government's helm? Cheating carmakers? The regulators who failed to catch them?

The truth is, there's plenty of blame to go around. And beyond the original mistake, what bothers me is the absence, for so long, of any action to reverse it. Instead of trying to solve the problem, David Cameron's government, then Theresa May's, put their energy into fighting off lawsuits demanding a serious pollution strategy, and seeking extensions to EU air quality mandates. As mayor, Boris Johnson delayed and diluted his predecessor's more aggressive cleanup plans, and even sprayed dust suppressants near pollution monitors to artificially lower readings.

One of the politicians engaging most seriously with the problem, trying to rectify the errors that have poisoned his city's air, is Sadiq Khan. Abroad, he's known mainly as London's first Muslim mayor. Indeed, his personal story is compelling, and he talks often about his upbringing as the son of a Pakistan-born bus driver. At a time when anti-immigrant, border-closing populism has surged through the country's politics, Khan's boyish visage has become the face of a different Britain, metropolitan and multicultural.

Since his election in 2016, Khan has rolled out a series of measures aimed at cleaning the city's air. The "T-charge"—it stands for toxicity—that he slapped on the very oldest vehicles entering the center of the capital the following year was the first significant step on dirty air in nearly a decade. "I refuse to stand by while Londoners are killed by pollution," the mayor wrote as it came into force.[24] He announced plans to tighten emission rules further within the charging zone and expand it to a larger area. He set out his vision of a pedestrianized Oxford Street, ordered upgrades to clean up buses, and sought the authority to restrict wood burning in the city (an unexpected pollution source we'll come to later). Khan also created a system of alerts for schools on the dirtiest days, posting the warnings at bus stops and Underground stations too.

Those steps are meaningful, and there's evidence they've delivered some improvement. But there's only so much a city can do to clean its own air. Sadiq Khan knows better than most that the powers of a nation—and a continent—must be brought to bear to make real change happen. He's called, in particular, for a national scrappage fund to help individuals and businesses, many of whom bought diesels with government's encouragement, to replace them. "I am using all the powers I have to their fullest extent," Khan has said. But "I have one hand tied behind my back" by inaction from above.[25]

Today, pressure for broader change is growing, as fear and anger bubble up in London and beyond. Britons—and Europeans more widely—have finally woken up to the diesel disaster unfolding around them. Air pollution is in the news now, even, sometimes, in the top-of-the-hour headlines when the radio wakes us, or on the front page of the paper. It has entered public consciousness, too, at long last. Any mention in the Facebook group for parents at Anna's school draws dozens of worried comments, and I read often these days about Londoners raising their voices to demand something better. A shift back to gasoline may be the likeliest first step, but it is long past time to think hard about what comes next, about finding fuels, and ways of getting around, that aren't just incrementally better, but truly clean.

I got a glimpse of a different future when roads in central London were closed one weekend for a light festival, and whimsical art installations dotted the urban landscape. My mom was visiting, and as we strolled with Anna down the middle of Regent Street, I marveled at the pedestrians thronging a space normally crowded with cars and buses. The grand facades of the buildings lining the road were suddenly visible in a new way, and the buzz of excited conversation replaced the rumbling of engines and honking of horns. And the air—wondrously—felt fresh, free of the fumes that normally cloud the center of town.

It's just a memory now. And there's a long way to go before Londoners can hope to breathe that easily outside of special occasions. My life here is a good one. Our friends and neighbors come from all over the world, the restaurants and theaters and museums offer boundless stimulation, and I've always found London more livable than many big cities, its parks and gardens providing frequent respite from busy streets. But the air bothers me, worries me, just about every day.

Whenever we can, my family and I walk along quiet roads rather than busy ones. I'm especially grateful that our house and Anna's school are connected by a woodsy urban walking trail we hurry along each morning and stroll in more leisurely fashion in the afternoon, with hardly a car in sight. I know such everyday choices make a difference. But the truth is, there's not much I can do about a problem that's so much bigger than any one person. At least this book, this writing, is an action. In a world that often seems determined to turn a blind eye to the dangers we inflict on ourselves, it's the only thing I know how to do.

On my way to school pickup one smoggy, sweltering summer afternoon, I heard an older woman just ahead of me telling two small boys they'd have to go straight home that day. "It's good to be outside," she said, "but sometimes the air is not good air."

4

AIR YOU CAN CHEW

Poland and the Price of Coal

They float in the sky, the invisible trio of atoms, three bound together into one. Carbon, and oxygen, and oxygen. Carbon dioxide. As dinosaurs' reptilian ancestors pick their way through vast swamps, the tiny molecule, one of uncountable billions, bumps against a leaf, is drawn in. Soon its atoms are wrenched apart, reshuffled, combine with other partners to become something new, part of a tall tree's body. But nothing lives forever, and the tree, eventually, falls. Sinks, as the weeks go by, the months, into the mud, deeper and deeper. Others fall on top of it, trunks and branches, giant ferns, mosses. And the years roll past, in their hundreds, their thousands. New life grows, new layers. Millions of years now, hundreds of millions. The ancient tree no longer has its old form, but what remains is rich with everything it once took from the air. The weight above presses down, compresses, and those remains grow dense. Solidify. Become, over eons, hard, and black. Then, suddenly, a sound breaks the silence. A rumbling, and a high whine as a drill pierces the earth. Before long, the work of millennia will be reversed, as everything nature pushed deep into the ground is brought back up, burned. And all it contains, the residue of time, of nature, is released back into the world. An atom of carbon, resurrected from its dark tomb, finds its old partners. And the trio, once more, is bound together, drifting upward.

One cold afternoon, in a quiet, leafy neighborhood a few miles from Krakow's medieval center, I walk past an empty playground, toward a pale orange and yellow house with a red tile roof. Inside, and up a flight of steps, it's warm and inviting. A kitchen opens up onto a bright space that fills many roles—dining room, living area, and bedroom to two young boys. Pictures of soccer stars are plastered on the ceiling that slants above their bunk bed, Spider-Man stickers adorn wooden cabinets, and Polish copies of the Harry Potter books sit atop a long table. While her sons bounce on a big plastic exercise ball, run around making airplane noises, then retreat to flip through magazines on the floor, a woman in jeans and a purple T-shirt stirs something on the stove, and a cat prowls around. Off to one side, pushed unobtrusively against a wall, is a slim white machine, a couple of feet high.

This is the home of Bogdan Achimescu and Monika Bielak. He, clad in dark corduroys and a long-sleeved black shirt, glasses pushed up onto his shaved head, is a teacher at the art academy. She, her brown hair pulled back in a ponytail, is a graphic designer. The rambunctious boys are Julek and Jasiek, nine and seven years old when I meet them. The family would fit comfortably in the funky, creative district of any British or American city.

The slim white machine is an air filter, the only visible hint that all may not be well here. The device, Achimescu tells me, is little match for the fog of coal smoke that poisons this neighborhood for more than half of every year. Even with the filter running, he has measured particle pollution far above recommended levels at the boys' bedside. Without it, the concentrations are higher than on a major highway. It is, he says angrily, "as if you would go to sleep in the middle of the street." Bielak pulls out her phone to show me a photo she took through the front window. In it, heavy black smoke pours from the chimney of the big peach-colored house across the street.

Krakow is a place that is, very literally, choking to death. Its air is among the worst in a nation that is home to more than two-thirds of Europe's 50 most polluted cities.[1] But this is also a place that has begun to push back, that is demanding to breathe cleaner air, insisting it has the right to do so. So Krakow's story is also one of hope. A story about what happens when citizens recognize a threat to their families, their communities, their nation, and decide to do something about it. And about how those who wield power—the power to decide what is news, or what the law should say—listen when people speak loudly enough.

In Bronowice, the pretty, well-off corner of town where Achimescu and Bielak live, things are particularly bad. Its single-family homes are less likely than the apartment towers in more densely populated neighborhoods to be connected to the community-wide systems of piped warmth known as district heating. And many prefer not to pay for natural gas. So, like millions of Poles in villages and towns across the country, like others across eastern Europe, and around the world, they store coal in their basements or backyard sheds, and burn it in low-tech home furnaces.

Julek and Jasiek, with their superhero stickers and sports magazines, are among those who must breathe the filth that results. For much of the year, "they basically don't go out," their father says. Ever since he had particulate levels measured at the playground across the street and found them six times over the recommended limit, he's forced the boys to give up afternoons climbing and running there. "During wintertime, I shuttle them between school and home at maximum speed, and we live here with the windows closed." Often, Bielak says, she doesn't even want to go out. "The air is stinking, it's horrible, it's dirty." All winter long, as smoke floats from their neighbors' chimneys, "we are like home prisoners."

Even so, the boys cough all the time, and every cold leads to a month of pained hacking. Sometimes, their mother says, it sounds like dogs barking. She pulls out a box and shows me the nebulizer—it's a device that turns medicine into an inhalable mist—she uses on Julek and Jasiek almost every night. When their breathing gets really bad, she adds steroids into the mix. Julek, the older boy, gets a daily injection of growth hormone too. His parents don't know whether his small stature has anything to do with the air, but it doesn't seem out of the question. "Everyone has to die one day, including them," Achimescu says. "But I'm wondering when is this going to happen, and how is this going to happen. And if it's not going to happen a lot sooner than normal."

It's not just the boys. Their parents feel the effects too. "We have these weird things, like I wake up in the morning one day and my nose is bleeding," Achimescu tells me. That lasted for 20 hours, so eventually he went to the emergency room. The doctor just laughed. "He says, 'What are you expecting? You're living in Krakow.'" The pulmonologist he saw for a nagging cough had a similar response. Bielak suffers from allergies that don't relent. Even the recently deceased family dog had a cough.

The couple worry incessantly, of course, and they burn with anger, too—at the neighbors whose furnaces poison their boys, at the government officials who have let the problem fester. And sometimes, even at one another. "Two years we've been quarreling," Bielak says. "Bogdan was having these plans, we should move to Berlin," while she wanted to stay, to live in the new home they've been building, equipped with more powerful filters.

Coal's defenders here often say it's used by those who simply can't afford cleaner ways of staying warm. To Achimescu and Bielak, that is a convenient myth that doesn't reflect the reality of their middle-class neighborhood. The family in the big house across the street built an addition a few years ago, but installed a new coal stove because they don't want to pay for gas. "There's a gynecologist nearby, she has a big SUV, and a big chimney that spits smoke," Achimescu says. Neighbors like that either don't understand the impact their decisions have on others, or don't care. In Achimescu's view, they might as well be chain-smoking right beside his boys. "Cancer for him," he says, pointing to one of his sons, "in exchange for a cheaper bill for him," gesturing across the street. "It's a great deal."[2]

* * *

Poland is one of the most coal-dependent nations on earth, drawing 85 percent of its power and more than 40 percent of its heat from the fuel.[3] So my visit here is a chance to see up close what coal does to the health of a country, a people, who rely on it. It is among the most toxic sources of energy, rife with poisons—mercury, arsenic, lead, sulfur, carbon monoxide—that not only taint air, but leach into water and are carried in the waste discarded by mines and the ash left behind by burning. Its smoke is thick with particulates, with NOx, with sulfur oxides that damage lungs and turn rain to acid.

This country's plight is just one piece of a story unfolding across eastern Europe and the former Soviet Union, in Bulgaria and the Balkans and Ukraine and beyond. Indeed, it is a story that resonates around the globe, from the power plants of India and China to the mines of Yorkshire and Appalachia. Coal provides more than 40 percent of the world's electricity,[4] and on top of the health problems it causes, it's a major driver of climate change, responsible for nearly half of all human-generated carbon dioxide emissions.[5] So the dangers Poles live with, the dangers that scar them a little more with each breath, are among the world's most existential.

In a yellow concrete building whose blocky, institutional style evokes the Communist past, Dr. Krzysztof Czarnobilski sees the cost every day. He's head of geriatrics at Krakow's MSWiA Hospital, a man with a formal manner and a neatly trimmed moustache, his white coat buttoned as he sits behind the polished wood desk in his office. Downstairs, workers hammer and shout as they renovate the emergency room. That's where Czarnobilski and his colleagues scramble to treat the stream of patients who show up whenever the clouds of black soot settle in over the city. Double the usual pneumonia cases arrive, the numbers of heart attacks and strokes jump, and those with breathing problems like chronic obstructive pulmonary disease and asthma come in struggling for air. "Ambulances queue here on this street," the doctor tells me.

Pollution shapes the lives of his elderly patients not just in moments of crisis, but with a grinding, everyday relentlessness, through much of each winter. Czarnobilski is in the uncomfortable position of advising them to stay indoors and avoid exercise when the air gets bad, a prescription whose bitter side effects he is well aware of. "The people are very sad on these days," he says. "They are lonely in their flats, and this depression is very, very harmful for their health." He's seen how the downward spiral can progress. The gloom and isolation get worse. "Of course, if an old patient stays at home and doesn't move," he says, before trailing off: "What else? Stay in bed, and finally he can die."[6]

I hear such anguish often here. Sometimes, it's tempered with a dark eastern European humor. "There is an old saying," one young father tells me, walking with his newborn daughter on a sunny Saturday. "The air in Krakow is very good, but you have to chew it."[7]

* * *

Krakow was once Poland's royal capital, home to kings, and it has the grand architecture to show for it. The historic center is graceful and elegant, a tourist's paradise dotted with domed churches, surrounded by thick stone walls. A fourteenth-century Gothic basilica towers over the spacious square at its heart, where I stroll past vendors selling the season's first tulips. It's all beautifully restored; money is a palpable presence in this part of town. Krakow is far more European, in the urbane, modern sense of that word, than I'd imagined. I realize as soon as I arrive that I'd been blinkered, as I conjured a picture of this place, by my journalistic focus on coal and a narrow set of preconceptions about Poland, centered around its wartime devastation and the heavy industry of the Communist years. I'd neglected to recall this city's rich history as a cultural and intellectual center, failed to anticipate its sophistication.

Bourgeois Krakow, filled with students and thinkers, didn't take to Communism, so the regime built factories and a huge steel mill to draw workers to the area, hoping demographic change would bring more fortuitous ideological alignment. Now, young people studying at some of the country's best universities live side by side with blue-collar families in the tall blocky apartment towers that are a short ride from the center on one of the shiny new trams that crisscross the city. Placards advertising day trips to Auschwitz are piercing reminders of the darkness in its past, but Krakow today is home to vegetarian cafés and hip pierogi places, upscale malls, and an old Jewish quarter transfigured into a nightlife and shopping hub, where visitors browsing hand-crafted souvenirs stop to peer into the shells of crumbling synagogues, faded photos of a lost past pinned to their walls.

In a pale orange building not far from the main square, three flights up a narrow staircase, a band of activists known as Krakow Smog Alarm were among the first to begin pushing for a law that holds the promise of real change. Their targets are those coal-burning furnaces, tens of thousands of them, that residents fire up every winter morning and evening. Coal companies often sell homeowners the cheapest, dirtiest grades, and the stoves are low-tech, their burning temperatures much cooler than in industrial-scale furnaces, so the worst pollutants escape into the surrounding air rather than being incinerated. Particularly fearsome is the potent carcinogen benzo(a)pyrene, present in the soot first identified in 1775 as the cause of the scrotal tumors afflicting the boys forced to climb, sometimes naked, into London's narrow, twisting chimneys, the young sweeps who emerged covered in coal tar and ash.[8] Today, in Poland's air, benzo(a)pyrene levels sometimes reach 10 or 15 times the legal limits, Krakow Smog Alarm says.

Unlike the smokestacks of power plants and factories, which—although they are killers, too—must at least meet filtration requirements and are subject to regulation, these residential chimneys were, until recently, almost entirely unmonitored. This is the smoke that makes Bogdan Achimescu and Monika Bielak's boys cough like barking dogs. So a few years ago, when Krakow, finally, decided it had had enough, when the street protests began, they were there, marching to demand better. Anna Dworakowska, one of Krakow Smog Alarm's founders, had helped plan those protests, a few hundred people at a time, many pushing empty strollers, stand-ins for the children forced, like Julek and Jasiek, to stay indoors. "There were notes saying, 'Sophie, she doesn't go out, she hasn't been in fresh air for two months now,'" Dworakowska tells me.[9]

Anna Dworakowska has short, spiky hair and piercing hazel eyes. Perched on the edge of a bright red sofa in a sunny office, she tells me how the campaign began, with a petition urging local leaders to ban the use of coal and wood for home heating. Many who signed left comments, too, and she and her colleagues invited artists to create drawings to illustrate some of them. One shows a man reaching for a noose that hangs from the sky; in another, the white eagle that is Poland's symbol wears a gas mask as it soars above church spires shrouded in smoke. A caption imagining the smog turning into a child's cancer is accompanied by a sketch of a cluster of buildings reflected in an IV bag. When the posters went up around the city, they got people talking. The local media, which for years had given pollution only cursory coverage, began to pay attention.

Soon, politicians followed where the public had led. In 2013, the local legislature passed a ban on home burning of coal and wood, the first in Poland, scheduled to take force five years later. Officials secured €100 million to help residents junk old stoves and install cleaner systems, whether gas, electric, or the piped-in district heat common in many parts of the country. By offering to shoulder a chunk of the conversion cost, but reducing the size of that incentive each year, they sought to encourage swift change. Although progress was slower than many had hoped, the effort continued even as the ban got tied up in legal fights.

Eventually, a court victory paved the way for a new version of the law, and Krakow's ban was set to take effect in 2019. There are other forms of pollution fouling the city's air, too, from the red-and-white-striped factory smokestacks that poke up around its edges to the cars that crowd roads, so the ban on home burning is only a first step. But in a nation that still holds coal in a tight embrace, it is undoubtedly groundbreaking. It will remove an immediate and egregious threat to the health of millions. And perhaps, in the long run, it will look like something even bigger, like the first hint of a broader shift, of a change that not only benefits Poles themselves but eases pressure on the planet, on the future, that belongs to us all.

* * *

Not everyone, though, is eager for change. On a bumpy cobblestone road, just past an ornate church, Arleta Wolek leads me down a half flight of steps to the basement of her small apartment building, where coal is piled on the bare floor of a boxlike room with dirty concrete walls. Pulling on a pair of yellow dishwashing gloves, she leans down to shovel chunks of the stuff into a blue metal bucket. One room over, she pours it into the hatch of an old furnace, its door caked with hardened black residue. Retired from her job as a manager at a cosmetics company, Wolek lives on her own, and this chore is just a mundane part of life. The trek downstairs, she says, sometimes provides the chance for a bit of socializing. Her brother lives in the building, too, and she often bumps into him in the coal room, where they chat while filling their furnaces and shoveling out ash.

It won't last much longer. Mindful of the financial help for those converting to cleaner heat, and aware that the amount on offer will soon dwindle, Wolek has decided to make the change, and is waiting for a workman to come install her new natural gas heating system. I suspect she doesn't realize it will be a boon to the health of the children I saw running through the heavy gates of the park down the block. She is worried about the cost. While she won't have to pay anything toward installation, she says she won't get any public help with her gas bill. There is assistance available, but perhaps she is too well-off to qualify.

Truth be told, Wolek is a bit puzzled by all the interest in heating systems. She seems particularly bemused by my request to watch this bit of her daily routine, and by the questions I pose through Marcin Krasnowolski, the scruffy, sweet young Pole who is helping with my reporting here. "Does she think it's exotic?" Wolek asks Marcin. He laughs as he translates her question. I can understand how she feels; I'm sure I'd react the same way if someone came to my house to watch me do laundry. But the fact is, I am intrigued. I've spent a few vacations in a Scottish island cottage with a coal fireplace, but other than that I can't think of a time I've come into physical contact with the fuel whose effects are so central to the issues I spend my days writing about.

For an American living in Britain, that's simply generational. When I share stories, back home, of my Polish travels, my English mother-in-law and New Jersey–born and –bred mom both respond by telling me about the coal that warmed their childhood homes. On Wainwright Street in Newark, my mother would watch, as a little girl, while a truck poured coal through a cellar door at the back of my grandparents' house. It's a useful reminder that it was not so long ago both my countries depended on the same fuel still sickening Poles. Marcin confesses good-naturedly that he, too, thinks it funny to meet someone who finds the sight of coal quite so novel. Growing up in Krakow, he used to carry it upstairs for his grandmother and parents, although he was glad when moving to a modern building in Warsaw, where coal furnaces are rarer, meant he could put that messy chore behind him.

While Arleta Wolek is resigned, reluctantly, to the coming ban, others are angrier. Marcin takes me to a lot beside some train tracks on the southern edge of town, where coal is piled on the ground in heaps taller than I am, some draped with blue plastic tarps, others open to the air. A worker with a bushy moustache trudges by with a pitchfork over his shoulder, and a compact white front loader pours load after load of coal into the bed of a dump truck. Next to a rusting railcar near the entrance, a banner bearing a picture of a smiling brunette advertises what's on sale here. "Quality is our priority," Marcin translates. "100 percent Polish coal."

This lot is Grzegorz Rumin's livelihood. He's been selling coal since 1998, and he knows the ban is likely to shut him down. For now, the customers are still trickling in, although the unusually warm temperatures mean this has been a slow winter. I can't help but wonder whether the global climate shifts caused in part by this very industry may be to blame. Every few minutes, another car pulls in and drives onto the beat-up weighing platform next to the little white hut that is Rumin's office. After they back up to a big pile, scoop coal into sacks, and load it into their trunks, customers get their cars weighed again and pay for the difference before heading home.

Rumin is a lanky man in his late thirties, in heavy boots and a frayed blue hoodie. Marcin relays my questions about the coal ban to him, but I don't need to understand Polish to know what he thinks of it. Standing outside the office, he is agitated and angry, the words pouring out of him. He gestures emphatically, then laughs in bitter exasperation. Rumin's phone rings incessantly, and he breaks off the interview several times to duck inside and enter an order in his notebook.

I don't know the Polish word for "bullshit," but it comes up repeatedly as Marcin begins to translate. I am reminded of the anger I heard half a world away, as a tired Delhi trucker pointed to a smoldering garbage fire and asked why he was the one who had to bear the burden of helping his neighbors breathe easier. A change like the one Krakow is making is costly. And as is so often the case when it comes to shifts that protect health and the environment, the losers know exactly who they are and what price they've paid, while the gains of the winners—in this case, the people of Krakow who will soon be free of the contaminants Grzegorz Rumin's coal puts into their air—are diffuse and less tangible, spread among millions.

With the number of coal stoves dwindling as cars proliferate, Rumin doesn't see why his industry should be targeted. He recalls politicians' boasts that the closure of a big local factory would improve air quality, but says he saw men lose jobs and their children turn to drugs and crime. "You can ban anything, but it doesn't make sense if you don't think about the consequences."

Marcin and I chat with some of Rumin's customers too. Adam Kukla, with big glasses and a mop of unruly white hair, has two heating systems at home, one powered by natural gas, the other by coal. Gas is far more convenient, but he usually burns coal. "Money is the only reason," he tells me. But there's something more than frugality at work, something less easily quantified. Kukla says a basement full of coal puts his mind at ease. "Sometimes the winters are very hard," and he worries for his children, his grandchildren. "It's important," he tells me, "to keep them warm." Another buyer, Zbigniew Jędrygas, hoists bulging bags of coal into an old Mercedes hatchback to stack in his garage for next year. His explanations echo Kukla's: This fuel's price is right. And more than that, "If I have one ton of coal, I feel secure, and I can sleep well."[10]

* * *

Those arguments resonate far beyond Grzegorz Rumin's coal lot, into the corridors of power in Warsaw where Poland's future is charted. Cost and security are the reasons political leaders give for their refusal to abandon the fuel that wrecks their people's health. They point next door to Germany, which, despite an ambitious renewable power revolution, still relies on coal, burning more than Poland does. If a nation with that kind of wealth needs this cheap fuel, I hear again and again, how can Poles, still struggling to emerge from the shadow of their traumatic history, be expected to afford better?

But it is the view to the east, not the west, that touches the deepest nerve in a country that has not forgotten what those long decades of subjugation to Moscow felt like. Poles remember, too, when Vladimir Putin turned off gas supplies to Ukraine in 2006 and again in 2009, and the suffering that followed as shortages rippled across eastern Europe. No one here wants to depend on Putin's pipelines, on his fickle good graces, to stay warm. Sometimes affectionately, sometimes sardonically, they call coal "Polish gold," the fuel this nation possesses in abundance, the one thing they can count on no matter what turn history takes next.

Then there's the belief that coal is crucial to Poland's economy, sustains jobs. In reality, the country's mining sector is so inefficient, its remaining coal so deep underground and expensive to extract, that the companies that do so lose more money with every ton they bring up. These days, Poland imports a good deal of its coal, and, funnily enough, one of its suppliers is Russia.

Coal is often cast, too, as a cherished part of Polish culture and history. Flanking the entrance of a technical college I visit are a pair of heroically proportioned, pick-wielding miners, sculpted in a blocky style reminiscent of old Soviet renderings of muscle-bound laborers. That image of strength and self-sufficiency underpins the deference the industry still gets, and its political power. Not long before my visit, angry miners responded with street protests and strikes to a government plan to close mines and restructure a failing state-run company. The prime minister quickly backed down. When European nations meet to discuss climate or air quality, to negotiate rules or deals that might help slow the planet's warming or ease its people's breathing, Poland is the eternal obstacle, coal's staunchest defender, delaying and diluting any effort to hamper its use of that precious black gold.

That stubborn attachment now stands in the way of a sea change that could help ease Poland's energy woes. It's not easy to transport natural gas long distances, so Poles can't just buy it from afar the way they can purchase Australian coal or Saudi oil. Russia is the dominant supplier in this part of the world, so it is true that relying on gas means dependence on Putin. A bigger-picture perspective, though, would take in the potential of renewable power—of solar, of wind—to fill that gap. It would add to the mix, perhaps, nuclear energy, or plug more rapidly into the international linkages that have begun to move electricity across Europe. But wind and solar are in their infancy in Poland, little encouraged by a government that seems utterly uninterested, if not downright hostile. Just as in the rest of the world, cleaning up energy supplies is, if not easy, eminently possible. It requires, however, an acknowledgment that change is needed, a plan to make it happen, and the political will—along with the money—to carry it out.

Most easily answered, perhaps, is the notion that coal is the best fuel Poland can afford. While the fistful of zlotys Grzegorz Rumin's customers hand over as they leave his lot is less than what they'd give the gas company, that tally fails to account for the steeper, hidden cost. The cost, for example, borne by the more than 45,000 Poles whom air pollution kills, every year, before their time.[11] In economic terms, those deaths cost the nation more than $100 billion annually, nearly 13 percent of its GDP.[12] "It's a myth in Poland that coal is cheap," says one expert I meet. "We pay for it a very high price."[13]

* * *

Coal furnaces like Arleta Wolek's are just one part of Poland's love affair with this fuel. So one morning, Marcin and I set out in a rental car to get a closer look at the power plants that are the other strand of the story. There's a thick, heavy haze in the air as we leave the city on a smooth new highway, heading into Silesia, the country's coal heartland. The landscape quickly turns rural, and it is infused with the dull grays and browns of a Polish winter, patches of snow on big, open fields. I glimpse a sign for Oświęcim and recognize, with a start, the Polish for a name that darkened the twentieth century: Auschwitz. Beyond an annual Hanukkah party, I'm not especially observant, but Judaism, Jewishness, is the faith, and the culture, in which I was raised. It was a secular, suburban sort of religion, but one that contained a sacred seriousness of purpose when it came to honoring the memory of Europe's murdered Jews. So I can't help but feel their ghosts hanging over the landscape here, and I'm glad recent decades have brought peace to this place.

We pass one coal-fired power plant after another, each with thick white steam billowing from its cylindrical cooling towers, and darker plumes of smoke wafting out of tall red-and-white-striped smoke-stacks. They sit beside roads, loom over villages.

Off the main highway, we wind through a pine forest, its tall, thin trunks bare except for tufts of dark green needles at their very tops. As we round a bend, two smokestacks poke up above the trees, and a moment later, we're parked beside the power plant to which they are attached. It's a hulking, pale green building, and a line of blue rail cars loaded with coal sits on snow-dusted tracks in front. Long metal pipes snake over a low barbed-wire fence, above the trains and toward the plant; so does a footbridge. There's no one around, and this isolated place feels almost otherworldly.

I'd hoped to get a look inside one of these plants, to see for myself how coal becomes electricity: the mills that pulverize endless tons of it into dust, the huge boilers, the spinning turbines. But, despite our many emails and phone calls, neither Marcin nor I has been able to persuade a power company to invite us in. So all I can do is gaze at this building from outside and snap some pictures with my phone before we drive away.

There's debate in Poland—I hear it even among environmentalists—about whether coal-fired power plants are part of the air pollution problem at all. The contribution of home stoves is so much more vis-ible, so egregious, that some suggest ending their use would solve this country's air woes. It would certainly help, but it's foolish to imagine burning coal for electricity is compatible with healthy air. Of course, the smoke those plants spew is more regulated, and better treated, than the stuff that wafts from families' furnaces, but it's far from healthy, and there's no doubt it adds to the burden of illness Poles bear.

And not just Poles. The pollution these behemoths create drifts across Europe, all the way to London, nearly 1,000 miles away. Indeed, one study found Polish coal-fired plants cause 4,700 annual deaths beyond Poland's borders. All coal smoke travels, but that's the highest toll any European country's power generation exacts on its neighbors. And of course, the accounting doesn't even begin to cover the impact on the climate.[14]

The industry's defenders like to note that power stations meet strict European Union emission rules, but in fact Poland has won postponements and exemptions to such requirements, so its bar has been lowered. Elsewhere in the world, others are confronting coal's dangers—Britain, for example, has promised to close all its coal plants by 2025—but here the burning continues unabated.

* * *

Karolina Zolna knows both domestic burning and the tall industrial smokestacks that jut up everywhere in Silesia are putting poison into the air she and her family breathe. We meet her in Myslowice, a small workaday town not far from the power plant in the woods. Another one, Elektrownia Jaworzno III, is visible from the main street.

Try as I might to focus on power stations in this phase of my travels, it's clear there's no getting away from those home stoves. A heavy plume of smoke floats from the chimney of nearly every house in sight. I see and smell it all around me as we speak in the biting cold, in front of a shop selling household goods like dishes and towels. The air is thick and heavy, astonishing in its awfulness.

Zolna is a school secretary, running errands with her son, whose eyes are obscured by a woolly blue hat pulled all the way down. She's warm and friendly, with purple eye shadow and a smart skirt, but there's a sadness in her smile. She says most of the children in this town, including her own two, have breathing troubles, struggling for air after even a bit of exercise. Her daughter was hospitalized with pneumonia for four months as a baby, and while the doctors never pinned down a cause, she has always blamed the pollution.

Stoves, Zolna points out, are used only in the colder months, but health problems here are year-round. "It has to be these power plants" too, she reasons.

She sees little hope for change on either front. "Let's face it, coal is the cheapest energy source," both for the nation and individual households. Her neighbors don't have money to spare, so as long as the current calculus holds, it's what they'll use.[15]

Bozena Borys shares Zolna's resigned anger, and her nagging anxiety. Her seven-year-old grandson, in Velcro-strapped boots and a fluffy hat shaped like a bear's head, fidgets and tugs her hand as she tells us she can see the Jaworzno plant from her window. Borys hates the smoke, is often short of breath, and says many in Myslowice have similar problems. She doesn't worry for herself: "At my age, it doesn't matter." The little boy lives in another town, also near a coal plant, and she hopes the air there is better than what she breathes.

But Karolina Zolna and Bozena Borys are the exception here. Many of those we meet in Myslowice seem unaware of the pollution, or at least unconcerned. When Marcin and I duck into some shops to ask about it, most of those we meet reply with shrugs or shakes of the head. "I've lived here since I was born, so I'm used to the smoke. I don't mind it," says a surly, red-aproned fishmonger. "I've never thought about it," the middle-aged woman behind the counter of a cramped liquor store tells us. Beside shelves stacked with juice, jams, and sauces, Agnieszka Kopec, a well-coiffed grocer, says the power companies tell neighbors their smokestacks use the latest high-tech scrubbers, and promise there is no danger. "We do believe them, because, well, we have to," she says. Many of her customers work in those facilities, and the pay is good. "So, I think it's OK."

A few steps off the main road, we walk toward a concrete apartment building with thick smoke pouring from its chimney. It's stinging and pungent, and quickly envelops us. An orange sign is tacked to the building's fence: "Coal in bags," Marcin reads. I sense something else too. I've heard about the plastic some Poles still burn in wintertime, bottles and shopping bags and assorted rubbish. Less commonly than in the past, and more surreptitiously, they pack the bottles with wood chips or shreds of old T-shirts, soaked sometimes in car oil, and toss a couple in the stove on a cold evening. They're known as grenades, and while most Poles now know how polluting they are, the heat they generate is both cheap and intense, so they still get used, another poison for this nation's lungs.[16]

Those I meet in the coal industry and the corridors of government often tell me of such burning. And while it is, of course, a real concern, it has also become a way to deflect attention and blame from the officially sanctioned contribution their own, far more widely used fuel makes.

* * *

A few days later, Marcin and I are on a train that takes us deeper into Silesia. With every mile, I see more clearly how coal shapes not just human lives, but a nation's landscape too. It sits in high piles beside the tracks, in the open-topped freight cars that pull up beside us, and in the rubble-strewn yards of crumbling industrial buildings. Coal is simply part of life here, visible everywhere. It all rushes by as I exchange selfies with Dan and Anna, on their way to school back home in London.

In Katowice, this region's commercial hub, Marcin has arranged for us to meet Andrzej, a slim, friendly miner in a camouflage baseball cap and gray sweat suit. As I squeeze in beside a child seat in the back of his little red car, we career toward his workplace. He parks at the foot of a towering hill that, in oddly straight lines, rises steeply, plateaus into a flat shelf, then rises again. This unnatural-looking mound, a good 15-minute climb to the top, is made of mine waste, Andrzej tells me (that is not his real name; he fears speaking openly would endanger his job). As we make our way up, I catch sight of the tall smokestack beside the mine shaft, and two towers of metal scaffolding, revolving wheels at their tops, that control the elevators hauling up coal and carrying men deep into the ground. There's a grayish warehouse-style building where the coal is cleaned and processed, and a series of towers beside it, linked by a half-dozen long, tilting chutes. Freight cars are parked along rail lines out front.

The buildings, I see as we ascend, sit beside a vast lot covered in coal, a carpet of deep, rich black, the size of two, maybe three, football fields. Bulldozers drive on top, pushing the coal into mounds, and big trucks laden with more make their way slowly up an access road. Every few minutes, another one drives onto the enormous pile and dumps its load. But as we climb farther, I realize I've been looking at only one piece of the whole. Behind the huge lot is even more coal, piled into hills that rise as high as 60 feet. This is product that can't be sold, Andrzej says, for Polish mining is a troubled, bloated industry, one being dragged down by its high costs. Everyone knows this business is dying; government subsidies can postpone the inevitable, but they can't stop it. Silesia's mines have been closing, one by one, for years. What's missing, for the most part—just as in hard-hit Appalachia and the northeast of England, the places whose discontent has roiled the politics of my two countries in recent years—is the support that could help Andrzej and others like him learn to do something else.

He hesitates when I ask what it's like at the bottom of the mine shaft. "I won't tell you how it feels. Everything is different, the taste of food is different," he says. "It's a different reality." The work is frightening, and while it has grown familiar, Andrzej is never complacent. As times have gotten hard, many experienced miners have retired, and working with a younger one is more dangerous. "Things can go wrong any time," he tells me. "You can get killed because he doesn't know what to do." The longer-term consequences are worrying too. Dust wreaks havoc on miners' lungs, and by 40, some are so short of breath they can barely climb stairs.

Andrzej's father and grandfather worked here before him. He's not quite sure how far back mining goes in the family history, "but I think many generations." Later on, back home, he pulls out a book on the industry's history—it stretches to the thirteenth century here—and points to a mention of his own mine, first opened in the 1600s. He's proud of that past, and although the job is hard, he knows he'd hate the nine-to-five of an office. Still, he hopes he'll be the last miner in the family line. His son, Antek, is just a baby, but Andrzej imagines a different future for him. When his own father was working underground, the paycheck was two or three times bigger than the average Pole's. Those days are gone, he says. "The history is dying."

From the top of the artificial hill, the view stretches for miles, everything shrouded in thick haze. I count four coal mines, each with a heavy black cloud coming from its smokestack, and a vast power plant, its own tall chimney beside four big cooling towers. There are houses, too, smoke floating from their chimneys and hanging over towns, because this is not a remote outpost in an unpopulated area. We are in the midst of a residential neighborhood, with apartment buildings next to the mine's entrance, homes across the street. The towering black piles I gaped at sit just beside a busy road. Back in the car, we wait at the parking lot's exit while a long train goes past, its open carriages piled high with coal. A stone's throw away, I see a small girl, no more than five, wrapped in a red winter jacket, waiting at a bus stop.

The next day I get a last reminder of the place coal occupies in Poland's psyche. In the center of town, a metal elevator tower like the one at Andrzej's mine stands on a low hill. But this tower is freshly painted, surrounded by sleek glass cubes that provide a window into the ground. What was once a working coal mine has been transformed into a museum of art and history. While the skeleton of the old mine has been carefully preserved, the place is also utterly reimagined, remade in the sleek, modern architectural vocabulary familiar to museumgoers throughout western Europe and the United States.

It's jarring to gaze into the sunlit underground space, where a team is busy installing video monitors, and think of the dust-covered men who once worked sweaty shifts here. The museum is a strange hybrid. While it is surely a testimony to the part coal has played in shaping this nation's history and identity, it also seems to communicate a desire to leave that past behind. Or just a recognition, perhaps, that it must eventually fade.

*　*　*

My trip ends in gray and sprawling Warsaw, utterly lacking Krakow's beauty, and looking very much the Communist capital it once was. But pulsating, beneath that dull surface, with modernity, with culture and creativity and the energy of a new generation of outward-looking young people like Marcin and his artsy friends. On a free afternoon, I ride a tram to the Warsaw Uprising Museum. My historical imagination constrained, perhaps, by my own background, and my limited knowledge of Poland's past, I'd anticipated an exhibition on the final days of the Warsaw Ghetto, the doomed, courageous last stand of this city's once-vast Jewish community. But the museum tells the story of a different rebellion, of the Poles who rose against a dying Nazi war machine and were crushed by a brutal bombardment, of a battle that killed hundreds of thousands and all but flattened a city, of betrayal by the Soviets who watched the decimation and waited to march in after the dust settled.

Browsing, not long after, in an airport shop, I spend my last few zlotys on a souvenir that evokes a brighter moment in this nation's painful past. A mug bearing the red-on-white logo of Solidarity—Solidarność, in Polish—it's a bit of kitsch, I suppose, a post-Communist, postmodern tribute to the shipbuilders' union that helped topple an empire. And I wonder, later, sipping coffee from that mug back home, whether Poland will find the inspiration, the courage, to transform itself once again. Whether we all will.

Today, with the winds of a dark, angry populism, of nationalism and isolation and disdain for science, blowing in Europe and the United States, in Warsaw and Washington and points between, it can feel naïve to hope we'll summon that collective will in time. If we do, though, and if we carry out the peaceful revolution needed to end our reliance on coal and the other fuels that are killing us and our planet, to leave them where they lie and build a cleaner, healthier world instead, that change would stand alongside the epochal year of 1989, when decaying regimes crumbled and a hated wall fell, as an equally proud, noble chapter in human history.

5

COWS, ALMONDS, ASTHMA

Crisis in the San Joaquin Valley

The body senses trouble and knows it must respond. Something alien has arrived, something threatening. A tiny particle, deep inside the lung. The soldiers that rush to fight it wield a weapon that's potent, but dangerous: a reaction powerful enough to destroy many invaders, but capable of turning against its master too. Able to leave this battlefield and journey through the body. Unleashed now, the cells rush from the lung, into the bloodstream. Still fighting their fight, wielding their weapon, but far now from the enemy for whom it was intended. The weapon that is at the root of so much that ails us: inflammation. Thickening the walls of arteries, so they grow a little narrower each day. Thickening the blood itself. Until, eventually, a clump forms from the thickness. A clot. Floating through the body until it catches somewhere, comes to rest. Blocking the flow of blood, of life. To the heart, perhaps, or the brain. The original invader is long gone now, but the cascade of consequences it set off is only just beginning.

Highway 99 cuts down the middle of the San Joaquin Valley, hot and dusty. The trucks that ceaselessly crowd this road hurtle by, or ride close on my tail, as I drive a behemoth of an old borrowed car southward, past billboards and motels and fast-food places. To the right, the Diablo Range spikes high into the sky, its peaks rising steeply from an utterly flat valley floor. On my left tower the majestic Sierra Nevada. In the parched strip of land wedged between the two ranges—home to impoverished immigrants and the vast agricultural concerns for whom they pick peaches and tomatoes and pistachios—the sun beats down relentlessly.

The highway is lined with the hulking architecture of industry. But it looks like no industry I've seen before. Between the stretches of open fields, cylindrical tanks, some of them four or five stories high, cluster together, pipes winding along their tops. Long enclosed ramps shoot from cavernous buildings. Train tracks run along either side of the road, their branches bringing rail cars right up to the plants. Farther back from the traffic, beneath white roofs held up by metal poles, cows cluster by the hundreds on concrete platforms.

This is industrial-scale agriculture—factory farming to its critics, or just Ag to locals. The San Joaquin forms the bulk of California's vast Central Valley, one of the world's most fertile and productive farming regions, providing a quarter of the food Americans eat.[1] The hulking structures here on 99 process the corn and soy shipped from the Midwest into cattle feed, dehydrate milk for export to China, and ready the crops grown in nearby fields for shipment far and wide.

But the San Joaquin Valley's agricultural productivity isn't the only quality that earns it superlatives. This is the worst air in America. The region's cities and towns consistently dominate the American Lung Association's particle pollution rankings, and only Los Angeles beats it for ozone.[2] Pollution is a tangible presence, shrouding the mountains in a perpetual haze. On some days, locals tell me, you can hardly see them.

The crisis here shatters notions that dirty air is only an urban problem. But while some of the causes of the San Joaquin Valley's pollution may feel unexpected, its story mirrors a much wider, very American one. So the parallels to places like Houston and Pittsburgh, New York and Cleveland, are stronger than the uniqueness of this landscape might seem to suggest.

One common feature is that despite decades of improvement, the threat still hanging in the Valley's air, and the country's, remains very real. Another similarity lies in the double-stranded dynamic at play not just in the San Joaquin Valley and across America, but around the world: While dirty air is in many ways an equal-opportunity problem, harming everyone who lives in a polluted region, it also tends to exact its steepest price from the poor. Disproportionately people of color or new immigrants, they live near the busiest highways, the dirtiest factories, and—in this place—the foulest-smelling concentrations of animals. So the story of air pollution, in the Valley and far beyond it, is threaded through with issues of race, class, and fairness.

The third thing the tale of the San Joaquin Valley's dirty air shares with the nation's is the bitterness of the arguments over what to do about it. The battle lines are nearly as old as America itself. Big business, and the politicians who serve its interests, standing off against demands that government wield its might to restrain corporate behavior, in the name of a broader public good. As ever, the fulcrum of the fight is the tool used to do so: the power of regulation.

* * *

I get off Highway 99 at an exit marked Shafter, onto a narrow two-lane road. The long rows of trees on either side are bare, and the ground that surrounds them is dry and dusty, like everything here. Later this year, these trees will bear almonds; the Central Valley produces 99 percent of America's supply of the nut[3] and 80 percent of the world's.[4]

Soon, I pull up to Tom Frantz's ranch house, set amid his own almond fields. Roses grow out front, near a pink Adirondack chair and a white picket fence; there's a parched dirt yard beside the driveway. Frantz grew up in this house, and he has a grizzled, weather-beaten face, the face of a man who's spent much of his life outdoors, a farmer, and also, until his retirement, a high school math teacher. A mane of tousled gray hair springs from his head, and he wears a frayed tan T-shirt and bushy moustache. Frantz is a lifelong gadfly, a relentless litigant and activist, head of a group he cofounded, the quirkily named Association of Irritated Residents, or AIR.

It's a name I'll hear, a week or so hence, in the hush of a high-ceilinged, marble-walled federal courtroom a couple of hundred miles away, in San Francisco, where Tom Frantz will cut an out-of-place figure amid the dark-suited lawyers and black-robed judges compelled to engage with his latest legal battle, against a big dairy near his home and the government regulators he accuses of enabling its polluting ways. For now, though, he's in his element as he leads me inside, through the laundry room, and takes a seat in a rolling office chair.

Here in Kern County, Frantz tells me, every square mile is home to thousands of cows. "They're wall to wall, almost." There's a spot near his house, he says, that's within a five-mile radius of ten mega-farms with a total of about 60,000 cattle.[5] Nearly a half million live in nearby Tulare County, more than a million and a half in the Valley as a whole.[6]

Frantz pulls up a satellite map that offers powerful evidence of their impact. It shows a yellow-orange circle centered over Highway 99, the color a warning of high levels of methane, the main ingredient in natural gas. In the United States, the only place with more airborne methane is the point where Arizona, New Mexico, Utah, and Colorado meet, and gas is extracted from coal beds.[7] Here, though, it doesn't come from mining: "There's no source of methane except the cow manure." Methane is a potent driver of climate change, so these huge concentrations of cattle pose a threat not just to their neighbors, but to the planet's very future.

Ammonia wafts from the big lagoons of excrement, too, and from fertilizer in the fields; it combines with other pollutants, including nitrogen oxides from vehicles, to create PM2.5, the deadly bane of this Valley.

Frantz clears up something that's been puzzling me. Dotted all around the Valley are huge piles—hills, really—as tall as houses and hundreds of feet long. They're covered with white plastic tarpaulins whose smooth surfaces are studded with big black circles, and I've been trying to fathom what's underneath. Frantz explains they're heaps of silage, fermenting green maize or other fodder for cattle. The black circles are sliced-up tires, there to hold the tarps in place. It turns out these odd-looking mounds add their own taint to the air here. When the tarps are lifted, gases called volatile organic compounds, or VOCs, escape, then combine with NOx in the air to form ozone.

The smell can be oppressive for those who live near the dairies, blighting everyday life and making many ashamed to have guests. On summer evenings when the air is still, "every one of those cows is kicking up dust with every step," and it drifts into homes, Frantz tells me. "It's manure that you're breathing, not even dirt." He suffers from bad headaches when particulate levels climb, and says that if he ventures anywhere near a dairy on such days, "it's ten times worse."

The effects of the Valley's terrible air reach well beyond those who live next to big farms. The region's premature birth rates are among the highest in the state,[8] a problem likely to be caused by a confluence of factors, although I know from Beate Ritz's work that it is linked directly to dirty air. At the other end of life, "I've watched people die prematurely. I will die prematurely living here in Kern County all my life," Frantz says matter-of-factly. He's certain that's why his father's life was cut short. While a heart valve problem had genetic roots, Frantz says dirty air hastened his dad's decline, and his death ultimately came during a winter of high particle pollution. "If he had lived on the coast, he would have lived easily 10 years longer, I don't have a doubt about that."

Then there's the quality of life: "You're sick a lot." He and his father both developed asthma as adults, and so did his brother-in-law. "I see my daughter getting migraines when she comes here to visit."[9]

Eventually, we head outside, and into Frantz's pickup truck. Just down the road, kids are playing outside the school he attended as a boy. When a big dairy opened up nearby, he tells me, the classrooms were besieged by flies, although there are fewer now that a grove of trees has grown up into a barrier.

The mega-dairies are relative newcomers. "None of those were here in 1990," Frantz says. While the Valley's fertile soil has long made it hospitable to growers, it was only as property values around L.A. began to skyrocket that the owners of that region's big milk operations sold up and moved to Tulare and Kern Counties.

As we pass one huge farm after another, Frantz explains that the open-sided platforms crowded with cows are called freestall barns. "Here's a dairy with one, two, three, four" of them, and a milking area in between. He guesses its permit covers 3,500 milking cows, which means that with calves and heifers and animals rotating through resting periods, there could be as many as 7,000 head in total. There's another farm, of similar size, right across the street. "I keep a map in my office that has the size of all these dairies," Frantz says, gesturing toward the biggest one, which he tells me is permitted for 6,000 milking cows, so probably houses 12,000 animals in total.

Soon, we bump onto a dirt road behind a dairy and get out beside a huge rectangular lagoon. Dealing with vast amounts of manure is the biggest challenge facing these farms, which are known in the industry as concentrated animal feeding operations, or CAFOs. In a more traditional, smaller-scale operation, manure becomes fertilizer, a rich source of nitrogen that nourishes feed crops, its nutrients moving from animal to soil to plant in an eons-old cycle. But with so many animals on so little land, there's far more manure than the soil can absorb, and much of the feed, in any case, is shipped in from afar. So workers put the excrement of thousands of animals through a separator, which removes solid matter that is little more than straw, leaving the nitrogen and other nutrients in liquid form. That's what fills this murky pool, which must be at least 100 yards long.

American companies invented this style of agribusiness, and it has reached its apotheosis here in the San Joaquin Valley: "The most industrialized farming in the history of man," says one local writer.[10] The scale is hard to fathom—a million acres of almond trees; a single company that, every week, grows enough carrots to circle the globe.[11]

Such intensive production has changed, in just a generation, our relationship with the land we live on and the food we eat. It's helped— through its heavy use of fertilizers, pesticides, and the medications necessary to prevent disease in animals living cheek by jowl—to drive the terrifying growth of antibiotic-resistant bacteria, the dwindling of insect populations, the impoverishment of soil. In this dry Valley, it's brought groundwater depletion so severe that in some places land has sunk nearly 30 feet since the 1920s.[12]

And, as I now understand, it is also a major contributor to air pollution, not only here but around the world. Agriculture, I'm shocked to learn, is responsible for about half the man-made air pollution in America and even more—55 percent—in Europe.[13] In much of Europe and the eastern United States, it's the largest single cause of air pollution–linked deaths.[14] In Britain, one study found European farms—not the Saharan dust that government and the media blamed at the time—were the main cause of one particularly awful pollution episode.[15]

The roads that take us through the almond groves back toward Frantz's house are a grid of straight lines and right angles. Black hoses, the final links of vast irrigation networks, snake through the dirt from tanks perched at the edges of fields. I notice, too, white wooden boxes scattered around the groves. They hold beehives trucked in from across the country, the only way to meet the demand for pollination that will come all at once, in a few weeks' time, when these trees begin to blossom. At one point, honeybees were flown here all the way from Australia. It's known as the planet's greatest annual "pollination event," and the millions of rented, out-of-town bees—stressed by travel, exposed as they work to alien microbes and parasites—are as vivid a sign as any of the ways this intensive, modern kind of farming has warped nature's rhythms, and placed living creatures under extraordinary pressure, all in an effort to wring maximum productivity from every inch of land, and every dollar put into it.

* * *

While the agricultural sources of the Valley's air pollution are new to me, the more familiar contributors are here too. The trucks that barrel along 99 and the parallel I-5, calling in at local farms and processing plants, are a big part of the problem. Diesel fumes also come from field equipment—tractors and such. All of it gets trapped by the towering mountains, unable to blow away.

And after saying goodbye to Tom Frantz, I see that farming is not the San Joaquin Valley's only business. Kern County is also the center of California's oil and gas industry, home to three-quarters of the state's fracking operations and conventional wells.[16] Amid the almond groves and alfalfa fields, big pumpjacks are scattered everywhere, their heads bobbing up and down as they pull fuel from the ground; flames lick up from the towers beside them.

Before getting back on the highway, I stop in a sleepy town called Wasco, where two older men in big cowboy hats sit on the front porch of a small house, looking, at least to my big-city eyes, like characters from an old Western. There are asphalt and concrete operations, and a big coal terminal, its huge cylindrical tanks connected by long chutes. Right across the street is a cluster of a few dozen bungalow-style buildings, home mainly to immigrant fieldworkers. Laundry hangs on clotheslines, and children toddle around a small playground.

It wasn't far from here, 80-odd years ago, that another new arrival—a threadbare, road-weary young man with a pack of empty-bellied relatives—searched desperately for work, found dignity and decency in a government-run migrant camp, and met, beyond its gates, the full force of humanity's darker side, its greed and hatred and coldheartedness. Tom Joad was fictional, of course, the protagonist of *The Grapes of Wrath*. But hundreds of thousands just like him, Okies fleeing drought and wrenching economic change, felt the same hardness when their jalopies staggered into this Valley in the 1930s. They picked peaches and cotton for starvation wages, and camped, like the Joads, in miserable shantytowns. John Steinbeck met some of those migrants when he traveled the San Joaquin Valley before publishing the novel that ripples with fury at their plight.

Dorothea Lange toted her camera through some of the same encampments, taking photographs that still define the Great Depression in America's historical memory. Her most enduring image, of a dark-haired woman, anxious and worn, staring into the distance as two children hide their faces behind her shoulders, was captured just outside the Valley, in Nipomo, near the coast. And a generation after Tom Joad hid from vengeful police doing the bidding of cruel landowners, and fled into the night to become, perhaps, a labor organizer, another icon of the forgotten, a nonfictional one, also fought for justice here. It was from Delano that Cesar Chavez and his United Farm Workers led the grape boycott that drew a nation's eyes to the exploitation of itinerant pickers in the 1960s.

All of which is to say that inequality, the painful chasm separating the comfortable from the barely surviving, has a long history in the San Joaquin Valley. It separates, in particular, those who tend the Valley's crops, who gather its tomatoes and pecans, who separate its manure and staff its factories, from those who reap the rewards of all that labor.

That inequality is woven into the DNA of the Valley's economy. Southern cotton men came here in the 1920s when the boll weevil destroyed their crop back home, and they brought their plantation mentality with them. Later, big growers recruited workers from Mexico and "basically imported a whole lower class," says author Mark Arax, who has chronicled the region's history. Even today, he recounted in a television interview, its structure is nearly feudal. "The land and the machines are controlled by maybe 300 families up and down this Valley."[17]

Those economic forces shape both the region's air pollution problem and residents' experience of it. Locals say just about any farm, factory, or recycling plant can get a permit to operate here, so the Valley is crowded with the facilities no one else wants. Incinerators burn branches and other woody detritus from farms, big cities ship their kitchen waste and sewage sludge to an industrial-scale composting plant, and oil comes in by train to be poured into pipelines headed to refineries on the coast.

The result has been a health and environmental disaster that goes well beyond the air. In a state database that aggregates the effects of many kinds of pollution, from pesticides and toxic waste to dirty air and water, the San Joaquin Valley is home to 7 of the 10 most burdened neighborhoods in California.[18] I hear about the box factory whose dust coats the cars, and the lungs, of its neighbors. And a rendering plant in Fresno, the Valley's biggest city—and one of America's poorest—where the stench of animal carcasses fouled one of the most deprived neighborhoods for years. It's a dynamic playing out across the country and around the world, familiar to those who live beside big garbage transfer stations in the South Bronx, to Chicagoans who fought to force huge stockpiles of dusty, toxic petcoke (a by-product of oil refining) out of their neighborhoods, to Londoners whose housing budgets consign them to the city's most traffic-choked corners.

José Chavez knows the story well. He lives in Arvin, where flush toilets at the "gov'ment" camp frightened Tom Joad's little sister Ruthie in *The Grapes of Wrath*; Steinbeck based his Weedpatch camp on the town's federal shelter. Arvin is still home to impoverished fieldworkers; today most are Latino. I meet Chavez only after I've left the San Joaquin Valley, several hundred miles away, at a park in Oakland. He's risen before dawn to pile onto a bus with several dozen neighbors who've come north to attend a protest calling for action on climate change. Someone pulls out a box full of burritos wrapped in tinfoil, and the group munches on an early lunch as they wait for the march to begin.

Chavez, an orange picker, wears a too-big gray vest over a short-sleeved white shirt, and his brown pants are baggy. His dark hair is graying at the temples, his eyes heavy-lidded, his face somber and serious. He doesn't speak much English, and my high school Spanish is rusty, so another marcher translates for us. Chavez is from Guanajuato, Mexico, I learn, but he lived in Los Angeles before moving a few years ago to Arvin, which sits in the area with the highest particulate levels in the Valley. Indeed, its PM2.5 concentrations may be the worst in the nation.[19]

That, he says, was when his health troubles started. He and his teenaged daughter both have asthma, developed since they arrived in the Valley. Chavez says his wife and another daughter have milder respiratory problems, and both the kids next door are asthmatic too. In fact, nearly everyone he knows in Arvin has health concerns of one kind or another; breathing difficulties are most common. The causes of the community's suffering are many. There are lots of dairies around, and their stench is awful. A big composting plant adds its odors. "I passed through there this morning," Chavez says. "It was really bad, you could smell it. I had to roll up the window."

The town has been cited for arsenic levels three times the legal limit in its drinking water.[20] Pesticides take their toll, too, and Chavez recalls being in the fields once as crop dusters swooped in overhead. Is he angry about all of this? I ask. Yes, comes the answer. That's why he joined the Committee for a Better Arvin, a residents' group that is fighting on many fronts to improve life for residents. "It's unjust, what they do to us."

Javier Cruz has come to Oakland for the march too, an older man in a big tan cowboy hat and gray suit. When I comment on his unusual belt—not just snakeskin, but an actual preserved snake, its head serving as the buckle—he laughs, then gamely unfastens it to show me he can even open and close the jaws. There's no one around to help with translation this time, so we must make do with my gringa Spanish (a more fluent friend will listen to my recording of the conversation later). Cruz, originally from Michoacán, in Mexico, lives now in Delano, just up Highway 99 from Tom Frantz's home, and his community is also besieged by polluters. There are the dairies, of course; chicken coops, too, and fracking operations. Cruz says illness is everywhere in Delano, from colds and tonsillitis to cancer. He's a cancer survivor himself, and recovering from a stroke. "All my problems are because of" the pollution, he tells me, and he generally tries to avoid going outside because the air is so foul. The water has an awful stench, too, so he and his neighbors have to buy bottled.

When I ask whether others understand as well as he does how profoundly all that pollution affects them, Cruz sounds sad. "People in Delano talk about this, but they know nobody will listen. So, eventually, they stop talking." Like José Chavez, he works with a local group trying to give voice to his community's pain. Few in power seem to want to hear about their problems, but he'll keep at it. "You know, if the president himself asked me to talk to him personally, I would tell him," he says, and I believe him. "I'm not afraid."

* * *

Of course, while the poor bear more than their share of the effects, everyone who lives in the Valley must breathe its awful air. Pollution shapes the rhythms of everyday life here, in cities like Fresno, Bakersfield, and Modesto as well as the farmland and small towns that surround them. Asthma is common, and when the air gets bad, schools keep pupils inside, canceling recess and PE. Balancing young people's need for exercise against the danger posed by pollution is an ongoing struggle. There's an emotionally charged debate over whether to cancel Friday-night high school football games when pollution spikes. Coaches rarely do so, but some parents and air activists think it's irresponsible to let teenaged athletes risk their health by inhaling all that contamination. An eye-catching 2008 study—things have improved somewhat since then—put the economic toll of the Valley's air troubles at $6 billion a year, or $1,600 for every resident.[21]

It's Seyed Sadredin who has the job of confronting the problem, reducing emissions so those who live here can breathe more easily. He is, when I visit, executive director of the San Joaquin Valley Air Pollution Control District, the regional regulator that shares responsibility for air quality with the state. Sadredin has been locked for years in an angry standoff with the Valley's environmentalists, who believe he and his staff are more interested in defending polluters than protecting public health.

That fight—in its fury, in the size of the gulf separating the two sides and the slim odds of their finding common ground any time soon—is a microcosm of the partisan warfare being waged on the national level. Indeed, Sadredin has testified before Congress, pushing for changes to federal regulation that would ease the pressure on his district, and others like it, to deliver cleaner air.

Businesslike in a smart suit, he is trailed by a klatch of aides when he takes a seat beside me at a conference table in his agency's offices, a low white building behind a gated parking lot. Sadredin's message is clear: The Valley's air is vastly better than it once was, its businesses are more tightly regulated than any in the country, and there's little more that can be reasonably asked of them, because today's problem is caused mostly by forces beyond their control, from the trucks that zoom through the region to the particulates that blow in from China and the Bay Area, and the unfortunate topography that traps the mess in place.

There's some truth in that. Pollution that's traveled from afar does contribute to local woes, the bowl-like topography is a real issue, and the complexity of the ways in which airborne pollutants combine with one another makes it hard to nail down exactly how much dairies, for example, contribute to high particulate levels—although their impact on both neighbors' quality of life and the planet's climate is inarguable. Yet, like Polish coal companies' frequent mentions of the foul smoke from burning plastic, this focus on external causes also has the effect of minimizing what is under the Air Pollution Control District's power, and Sadredin's.

Environmentalists were incensed by comments he made to a local conservative radio host after the Environmental Protection Agency, during the Obama era, leveled a $29 million fine against the Valley for exceeding ozone standards, a fee local officials passed on to drivers rather than business. "The guilty party in that case is the federal government," Sadredin said on air. "Our businesses in San Joaquin Valley have spent billions of dollars over the years trying to comply with some of these outrageous regulations." When the host slammed national air quality rules as impractical, Sadredin eagerly agreed with him. "You're preaching to the choir," he said. "I'm hoping you and your listeners will join us to compel EPA" to stop pushing the Valley to take tougher action on polluters.

Sadredin's language isn't quite as strong when I meet him, but his bottom line hasn't changed. "We've already thrown the kitchen sink" at the Valley's dirty air, he says. "There's not much left that we haven't done." I also speak with one of the agency's staff scientists, a genial guy who, I was sorry to read, died not long after we met.[22] He told me of his frustration with local environmental groups he saw as fixated on the idea of Big Ag as a villain, an ideological hangover, in his view, from the 1970s battles to improve working conditions for migrant laborers. "They're kind of stuck with their standard bête noires—us and agriculture."

When, the next day, I meet with two activists from a group he'd singled out—the Center on Race, Poverty & the Environment—they laugh at the depiction of themselves as radicals from a bygone era. But there's no disguising the rift between the regulators responsible for the quality of this Valley's air and the groups pushing them to do more.

A faint smell of manure hangs over the farming town where the activists' offices occupy a little house with a trim garden out front. A shaggy black-and-white dog naps on the floor, and Lupe Martinez, the center's assistant director, gets up from his desk to greet me. He's low to the ground, a stocky man with a full, graying goatee and a wide-brimmed straw hat. Caroline Farrell, in a rumpled lavender shirt, is his boss, the group's director.

Their frustration with the Air Pollution Control District is clear. It is, in their view, forever looking for reasons to do nothing. "If there's a way of carving out a loophole," the agency does it, Farrell tells me.

"They will essentially say to industry, 'Tell us what you need.' That's the relationship." The reluctant regulators "try to discredit us, they try to pigeonhole us." But her center isn't against agriculture, she says; it just wants farms and other businesses to behave responsibly. "The people who were advocating for stop signs when we first got cars weren't anti-car. They wanted safety. It's the same thing."

Martinez, who lifts his big hat now and then to run a hand over his hair, says he's sometimes stunned by the health problems he sees among farmworkers' families. "I cannot get over the fact that we have one-year-olds, two-year-olds with the inhalers" to treat asthma. Those communities devote most of their focus and energy to making a living, but they are increasingly coming to understand the threat pollution poses, Martinez believes. His hope is that eventually they'll win a say in the decisions that shape their lives, and he does what he can to bring that day about.

* * *

John Capitman sits in the space between the advocates and the Air Pollution Control District. He's a public health professor at California State University, Fresno, and he serves on the agency's board. He got his seat there as a result of environmentalists' fight to gain a voice in the rule-making process, but he can see the business owners' point of view too.

An East Coast transplant, Capitman speaks in rapid-fire bursts, hunching forward to lean into each new thought. "This is a very polarized place," he tells me. Politically, it has "an almost colonial structure. There is a very small, very affluent group of people who run the Valley." Not surprisingly, they're the same big growers who own most of its land. For them, "it's always been about protecting the industry," and they give the area's politics a conservative, anti-government bent. That elite's vehement opposition to regulation sets the tone on the air agency's board. And engaging them in efforts to clean up, to change the way they do business, isn't easy, says Capitman. Are they interested in thinking about such questions? I ask. "No," he answers emphatically. "Why should they be? They're hugely profitable."

Even so, he also detects in his fellow board members a desire to live in a place that doesn't forever bear the embarrassing distinction of being the country's most polluted. So, while the district's actions have been less aggressive than necessary, more tepid than he'd like, Capitman believes it is, slowly but steadily, making headway. Among other signs of progress, he points to a big reduction in the burning of agricultural waste—not just uprooted nut trees and dead grapevines, but the bits of plastic and chemically treated wood discarded along with them. A deadly source of pollution, less common now than a decade ago.

As for his own role, Capitman tells me, "I listen a lot." Before every board meeting, he gets hundreds of phone calls. "The dairy council, the nut council, the environmental groups," he says. "I take the calls. I think it's the job." Ultimately, he believes, that kind of communication—and the involvement of the big growers—is necessary if the San Joaquin Valley's future is going to be different than its past. "There has to be enough of a relationship to work with the guy with 10,000 acres of carrots to get him to make that change, or the guy with 60,000 acres of grapes."

For now, there's not much incentive for the region's biggest companies to do things differently. Until that changes, it's clear who's going to do most of the suffering in a place still defined—decades after John Steinbeck wrote its story into the American canon—by its stark inequality. The same people who always have.

Late one evening, I find a bustling taco place for dinner. The clientele looks more like the people who work in the San Joaquin Valley's fields than the ones who own them. Tired from a long day of driving and talking, I watch the locals around me laugh and chat over the restaurant's cheap, delicious fare, enchiladas and shrimp tacos, as I devour a plate of guacamole and chips. Kids in strollers, weary parents, workmen still dirty from their labors, many speaking Spanish. An evening's pause, a meal, a rest, before work begins anew tomorrow.

6

HOME FIRES BURNING

A Paradigm Shifts

The smoke rises from thousands of fires, hundreds of thousands. Millions. Soot. Black carbon. The same particles that sicken the body, now heading toward the sky. Their color, dark, draws in the sunshine, absorbs the heat. The effects, in an atmosphere already tipping out of balance, powerful enough to cause clouds to form and to disperse them. To alter the weather. Bringing drought and flood. Hunger and thirst. But the soot doesn't stay long in the sky. Rain washes it down, or snow. It finds its way, some of it, onto vast glaciers. Glaciers whose runoff sustains life. Glaciers holding water away from ever-rising oceans. Their icy whiteness repels the sun, keeps them cold. Or it did, once. Now, the dark soot attracts heat instead. The ice melting, drip by drip. Shrinking.

Every day, three times a day—breakfast, lunch, and dinner—in hundreds of millions of homes around the world, women and girls build fires to cook for their families. They burn wood, charcoal, animal dung, the stalks and husks collected from fields after harvest. What they cook depends on where they live: lentils and chapatis in India, injera bread in Ethiopia, tortillas in Peru. The constant is the fires' thick, awful smoke. And the health hazard it creates: one even more deadly—because it's so much more concentrated—than the pollution from tailpipes and smokestacks.

More than 40 percent of humanity—33 billion people—lives in households that cook this way.[1] As economic growth has lifted many from the direst poverty, and as villagers everywhere have migrated to cities, that percentage has declined. But because the global population has been rising at the same time, the absolute number of people living with the smoke of open cooking fires remains stubbornly high.

Scientists used to call this "indoor air pollution," to distinguish it from the more familiar "outdoor" sources. But the jargon has recently changed, reflecting a new realization: that the two problems are deeply intertwined, because the smoke of cooking fires floats from kitchens and adds to an entire region's, or a nation's, dirty air. That the problem can't be solved just by venting smoke from homes, and that the cooks and their families don't escape it when they step outside—it follows them through their neighborhoods, through their days and their lives.

So the smoke from those home fires is now known as "household air pollution," while the broader problem once categorized as "outdoor" has been rechristened "ambient" pollution. And households' contribution, it turns out, is responsible for nearly 4 million deaths a year, about half the total air pollution toll. It's nearly 10 times as many as die from malaria. Because women and children tend to be home more than men, they suffer disproportionately. In addition to heart disease and severe, often fatal, lung ailments, the smoke can cause stillbirths and low birth weight, and it may also be linked to vision troubles like cataracts.[2]

It's a problem of poverty, of course. And a lack of access to cleaner alternatives, the ones we in the developed world take for granted: gas and electricity. Often it's a problem of physical distance from connections to those fuels: a rural problem. Ideas on what to do about it are shifting rapidly. While they do, billions of people await a solution, in desperate need of something better, their lives and health under threat until they get it.

* * *

Despite all that, cooking smoke is not Godknows Maseko's biggest concern. He and his wife, Helen, are responsible for the well-being of 64 neglected, abandoned, or orphaned children and young people. Before the Masekos took them in, most were living on the streets of Blantyre, the commercial hub of Malawi, a landlocked southern African nation that is among the world's poorest. The Masekos' days overflow with the work of managing their enormous household, keeping an eye on the kids' schooling, and running the small businesses that bring in needed cash—selling chickens and eggs, making clothes by hand, and renting out their minivan and two SUVs.

Maseko first came to understand how smoke from the wood and charcoal Helen cooks over might be affecting them and their children when he helped a British research team get T-shirts bearing the name of their project, the Cooking and Pneumonia Study, in Chichewa, one of Malawi's main languages. When he asked the lead scientist what it was all about, he began to realize the fires that had always been in the background of his life—for Helen, they're closer to the forefront— might be responsible for bouts of coughing and pneumonia among the dozens of children they regard as their own. "It's only now I'm getting older that I'm learning the dangers," he tells me when we speak on the phone. The smoke "is hazardous to the life of a person."

In Malawi, pneumonia, often caused by cooking fires, is the single biggest killer of children under five. They can languish for months, drained of energy, struggling to breathe. Maseko says that while the disease is common, parents sometimes don't recognize its seriousness, so they wait too long to seek medical care. He and Helen take their children to the hospital at the first sign of illness, but the doctors never suggest their coughing might have anything to do with the smoke that pours from the family's kitchen shed.

Even so, the couple knew they didn't want all those fumes in their house, so they built the shed, where Helen oversees cooking of the maize flour staple called *nsima*, along with beans, smoked fish, and chicken or beef stew. Two women come in to help, and the children pitch in too. In a photo on the family's fundraising website,[3] a couple of boys stand in a smoky corner, next to a tall black pot that rests atop three stones, a fire burning between them. It's a common arrangement: more than 90 percent of Malawian households cook this way.

The Masekos buy their fuel from vendors who bring it in from the countryside; villagers make the charcoal by heating timber inside air-tight kilns. In addition to the health problems they cause, wood and charcoal use are wrecking Malawi's forests. But better options are out of reach. The family has a friend with a gas stove, but he can afford to use it only one or two weeks a month. Gas is expensive, and so is the unreliable electricity supply; the Masekos sometimes have power for only seven or eight hours a day. The cost of solar panels is prohibitive too. The story is similar across sub-Saharan Africa, where 80 percent of the population—more than 700 million people—relies on the plant- and animal-based fuels experts call "biomass."[4]

After helping with the T-shirts, Godknows Maseko got more involved with the British study, even bringing the scientists to his grandmother's village. The researchers, based at the Liverpool School of Tropical Medicine, distributed thousands of metal stoves in which the traditional fuels could be burned more cleanly and efficiently. The villagers who got them would still be using wood and charcoal, but in a better, more modern device. For two years, the team tracked pneumonia cases among children. They hoped to see fewer in families with the new stoves, a sign the devices offered a healthier way to cook. Unfortunately, it didn't turn out that way. Women found the metal stoves easy to use, and they preferred them to the old open fires. But the pneumonia numbers didn't budge.[5]

* * *

While Africa is where the largest proportion of people burn biomass, India, with a smaller fraction—about 60 percent—but a vast population, has the highest absolute number: nearly 800 million people, in 170 million households, relying on smoky cooking fuels.[6] As with so many things in this enormous nation, the sheer scale of the problem can seem overwhelming. So the ambition of efforts to tackle it must be big too.

I get a close-up look at life with dirty cooking fuels—and a hint of some solutions—in a sleepy village an hour and a half outside the northern city of Lucknow. The trip there gives me my first glimpse of rural India: famously impoverished, deeply traditional, the left-behind home of so many of the migrants I saw sleeping on Delhi's sidewalks, doing its manual labor and driving its auto-rickshaws.

With my translator Neha and a guide from The Energy and Resources Institute (TERI), the environmental group that's helped us arrange the visit, I stop on the way at a small roadside restaurant. Tea is served in little clay cups, milky, spicy, and sweet, and we gobble down breakfast parathas with yogurt. Soon after, we're walking down a narrow, rutted dirt lane, past a woman in a sunshine-yellow sari who prods along a group of mud-encrusted pigs with a bamboo stick. Small homes line the path, some made of brick and some of mud, many with thatched roofs of straw and sticks. Children smile and shout to us, then pose for photos and laugh at the images on my phone. There are dozens of goats and cows around, their faces buried deep in brick or concrete troughs. Only later do I learn the name of this village, and scrawl it into a notebook: Mustafabad.

Walking into the Kushwaha family's home, we find ourselves in a courtyard-like space, open to the sky everywhere other than around its edges, which are covered by a bit of roof. Even so, I am immediately assaulted by thick, choking smoke that fills my throat and chest. My eyes burn, and it's hard to breathe; I begin to cough, squinting and blinking hard. It's like stepping onto the wrong side of a fire; the urgent instinct is to move away, to seek air that doesn't hurt. Instead, Neha and I begin to look around.

In two corners, the whitewashed mud walls are blackened, and in each a teenager or young woman squats beside a rough-hewn clay stove that is no more than a vessel for open flames. All their work is done at ground level; the stoves, low towers open on one side and on top, aren't even knee-high. A long stick pokes out from each one, its end aflame inside, the rest waiting to be fed in.

With smoke wafting up steadily, one of the girls kneads dough into circles and places them in the fire. Dressed all in white, her hair twisted up in a rather elegant chignon, she winces and rubs her face against her knees, then lets out a quiet groan and squeezes her eyes before returning to work. The breads puff up in the heat, then flatten when she pulls them out with metal tongs and piles them onto a tray. In the opposite corner, the other cook, her cousin, clad in turquoise, blows into the fire through a metal pole, coaxing the flames upward. A pile of kindling sits beside her, and ash spills onto the floor.

As they work, children wander in and out, and I take in the rest of this place. An older woman, a grandmother perhaps, squats on the ground too, scrubbing pots and pans in a stone-enclosed washing space, and a third teen bends low to sweep the hard dirt floor with a rough broom. Towels and saris hang from a clothesline stretched across the courtyard, and through curtains covering the entrance to the home's internal space, I can see a motorcycle. Two baby goats wander around the courtyard, and a handmade ladder leads up to the roof, a boy of about eight perched at the top.

The girl in white hurts her finger, and everyone fusses as she goes to wash it; a cousin takes over her stove. I'm shown to a seat, and while the men lounge outside smoking, the young cooks, and the other female Kushwahas, sit down, one by one, on a battered chair, to talk. An hour passes, then another, and slowly, I begin to piece together who is who, and to discern the outlines of their lives in this village.

All the while, as the young women speak and I wait for Neha's translation, my attention drifts to a girl hovering at the edge of the group, watching and listening. Her thick hair is pulled back in a pink plastic clip, and her bare feet are dusty beneath a dirty school uniform, the hem on one leg resewn with mismatched thread where it has ripped. She giggles when we catch each other's gaze, and a shy smile fills her whole face; her eyes are bright, curious. She has a thick, phlegmy cough, though, and I can hear the rasping beneath her laugh.

The older girls, I learn, are her big sister and cousins; some little brothers play nearby. The setup here is typical for India: two adult brothers and their families sharing a home, their wives and children all living communally, 15 people in total. The girl, I learn later, though she doesn't look more than 8 or 9, is 11. Close in age to my own daughter, and I suppose that connection is what draws me to Khushboo. The similarities I can see so clearly in two children who will never meet, both warm and kind and intelligent, the contours of their lives so very different—one filled with possibility, with every chance to learn and thrive and succeed, the other burdened by poverty, domestic work, and the brutal constraints rural India imposes on its girls.

* * *

I realize I have been thinking of the young Kushwaha women by the colors they are wearing—the two cooking when I arrived in creamy white and turquoise, the third wrapped in a blue and purple scarf. Now I learn their names, hear their voices. It's Savita, the third cousin, who speaks first. She's 19, tall, with an angular face, a hoop nose ring, wide eyes, and a broad smile. Confident and poised, she tells me she rises each day at 4:00 a.m., while her brothers and father are still asleep. She and the other girls make trip after trip to a well 10 minutes' walk away, fetching water for cooking, cleaning, and bathing. Then they squat at their stoves to make tea and bake bread for the whole family, each girl cooking for a half-dozen relatives, all in need of large quantities to keep their strength up as they labor in the fields where the family grows wheat, potatoes, and rice.

Savita is the only girl in her immediate family, so she shoulders the burden of food preparation, with help from her mother. All told, she spends six hours a day over the smoky fire, in addition to the cleaning, the water hauling, and work in the fields. I'm surprised to hear she's also working toward a bachelor's degree, through a correspondence course of sorts; she travels once a month to a college 20 miles away and studies at home during the weeks between.

But it's the cooking that defines her days. She's been doing it since she was 16. The boys bring wood from the forest nearby, and the stoves burn dung cakes, too, dried-out patties of cow droppings mixed with rice husks or straw. In India, these biomass fuels are known colloquially as *chulha*. Like everyone here, Savita aspires to a different fuel—LPG, or liquefied petroleum gas. It comes in canisters that only a lucky few in this village can afford, and it's far beyond the Kushwahas' reach, especially after unseasonable rains ruined much of their crop.

Such a stove would rid their home of the horrendous smoke, but nearly as important to the young cooks is its convenience, the ease with which it can be switched on and off, the hours of effort it would save. The traditional stove is laborious to prepare and start, and its flames must be coaxed back whenever they ebb. The girls often get burned, and Savita pulls up her sleeve to show me a long, thin scar. "Sometimes, I even tell my mother I wish I didn't have to cook," she says. She hopes she'll be married into a family with a gas stove, so the domestic drudgery of her adult life will at least come without the awful smoke of her adolescence.

Savita's cousin Renu, the girl in turquoise, is just 17. Renu is Khush-boo's older sister, and her voice is hoarse and low. As she speaks, she fingers a crystal that hangs on a string around her neck. She, too, cooks for hours every day, all three meals, but she gets less relief than Savita, because her mother is ailing, and struggles to take her place at the stove. "Peel, chop, make the spices," ready the stove before cooking. "I have younger brothers, they keep asking for more and more bread." Her hands are battered and dry, like those of a much older woman. But for just a moment, when she leans forward, she looks her age, young and unburdened.

Renu says her face is often covered with soot when she finishes cooking. Her eyes water constantly, and blowing to kindle the flames sets off fits of coughing. The headache never goes away. "I have a head-ache even right now, dizziness. I feel like my head is spinning all the time." She worries about what will happen when she and Khushboo marry and leave home. "If there's no gas stove by then, my mother will have to cook on this again," she says, gesturing toward the smoky corner.

The smoke carries an emotional sting, too—the shame of poverty, of being unable to afford something better. "If anybody sees me while I'm cooking, they'll wonder, 'Does this family not have enough money to buy a gas stove for the girls?'" she says. Most of her neighbors also cook with *chulha*, but not her big-city relatives. "They don't like to come here because there is so much smoke."

It takes a toll, too, on her learning. Renu has finished the first phase of her secondary education, and recently took the exams she hopes will win her admission to the second half of high school. But she can only make it to class once or twice a week because her mother needs her at home. "I get scolded by my teacher for missing so much school," she says. She wants to study medicine, but knows that's probably unre-alistic. So a neighbor teaches her sewing every afternoon. "I'll have to do something, after all, if my dream of becoming a doctor doesn't come true."

Renu's mother, Meena Kushwaha, is in her mid-forties, around my age, but she has the kinds of health problems I associate with the elderly. She blames two and a half decades of *chulha* smoke, and worries about what it's doing to her five children: "They end up breathing in the smoke." But it's hard to imagine accumulating the money for a gas stove. "We don't have anything left at the end of the month," she says. "How are we going to save when we can't even meet our expenses? We have to buy wheat flour for the house, we have to buy food for the livestock, education." They don't have enough to get Renu married, either, but Meena Kushwaha clings to the same hope for her daughter that Savita expressed for herself: in-laws with the means to have their daughters-in-law use LPG to cook for them.

As a young boy comes over and deposits a goat in Meena's lap, I ask about Khushboo. She's been coughing for two weeks, the mother says. The child's feet hardly touch the floor when she sits in the chair next to me, and she smiles broadly when someone snaps a photo of us. She's still too young to cook, but she grinds spices for the older girls, fetches water, washes dishes. At school, she studies math and Hindi, English too. She likes it there: "You don't have to do housework at school," she tells me. "I don't want to do housework when I grow up."

At home, Khushboo says, she starts coughing as soon as the fires are lit, and her eyes and lungs burn. She leaves when she can, spending hours beneath her favorite neem tree. "I just sit outside," she tells me. "I only come back when the smoke is gone." That escape won't be available much longer. Soon, she'll be old enough to start cooking. And like the others, she hopes for a gas stove. She knows her family may never be able to afford one, and if she must use *chulha*, "I will learn how to do it," she says bravely. "I'll ask my mother to teach me." But if a modern stove came tomorrow, "I would gladly cook now."

She has bigger dreams, too, of course. Neha asks what she wants to be when she grows up, and Khushboo looks down, playing with the weave of her chair and giggling a bit before answering: a teacher. I don't have the context to know how attainable that goal is. But I can see the life Khushboo's mother lives.

India is changing, even rural India, so perhaps these girls will become the teachers and doctors of their dreams. But change can be slow, and it's also possible their lives won't be so different from Meena

Kushwaha's, old before her time, worn to the bone by a life of toil and limitations.

* * *

As I leave the Kushwahas' house, I begin to realize that, just as coal shaped the landscape I saw in Poland, the evidence of *chulha*, of biomass fuels, is all around me here. Piles of twigs and branches sit in front of houses, and there's a towering stack in a little square where the lane opens out. I understand now that the cows crowding the village are a source not just of milk and yogurt, but fuel too. On cobblestoned ground and atop roofs, the cakes made from their dung are lined up in neat rows, drying in the sun. I'll see them, too, from the windows of a train I take for a weekend of sightseeing in Amritsar, home to an extraordinary golden temple sacred to Sikhs. As an attendant serves tea and biscuits and I send silly photos to Dan and Anna back home, I gaze out at pyramids of dung cakes, like haystacks.

Not everyone in Mustafabad has to live with *chulha*'s filth, so my visit to the village also provides a chance to see the beginning of a change, in all its complexity. Priti Kushwaha is a neighbor of Meena, Renu, and Savita, and her home is just a moment's walk from theirs. She's poor too, but I can see—and feel, in the air that doesn't assault me when I enter her home—that she's far better off.

The house is brick, and in its cool darkness, a few dozen electric lanterns sit on a shelf; she charges them with a bank of solar panels, and rents them out each night to neighbors who lack electricity. TERI, the group that has led me and Neha to this village, helped Priti Kushwaha start the lantern business a few years ago, and it's changed her family's life, bringing previously unattainable luxuries within their grasp.

A small woman in a patterned sari, she leads us up a narrow flight of steps to the roof, where a brick shed serves as her kitchen. Inside, it's clean and comfortable, with a concrete floor and a curtain rippling on an open window. Off to the side, I see a clay *chulha* stove like the one her poorer neighbors use, but these days, Priti Kushwaha rarely fires it up. Most often, she uses a higher-tech metal version, not unlike the ones whose benefits the Liverpudlians found so disappointing in

Malawi. She bought it from TERI, which is one of many groups trying to give villagers safer ways of burning traditional fuels.

Perched on a plank laid on the floor, she lights it up quickly to make us tea, placing sticks and straw into the insulated burning chamber, then turning on a fan inside the stove. Although the fuels are the same as those used down the street, this kitchen is far more pleasant than the other Kushwahas' courtyard (although they share a last name, the two families are not related). The metal stove consumes fuel more completely than the old-fashioned ones, leaving fewer unincinerated particles; its fan improves efficiency by pushing in air. Devices like this are known as "improved" biomass stoves, and they're undoubtedly less smoky than open burning, but the Malawi study added to a growing body of evidence that the improvement isn't enough to make much difference to their users' health.

Priti Kushwaha also has a gas stove, a two-burner setup that seems less like the kitchen ranges I know from home and more akin to the portable kind well-equipped American or British friends might bring on a camping trip. It's attached to a red cylinder a couple of feet tall. After hearing Khushboo's family's dreams of cooking with a stove like this, I'm puzzled when Priti tells me she rarely uses hers.

It turns out the gas is not only expensive, but hard to get. The shop that sells it is about six miles away, and I struggle to imagine how she or her husband carries the cylinder on their motorcycle. Even worse, she explains, "you stand the whole day in a line" waiting to buy the gas, and sometimes have to make a second visit. Then there's the cost, which comes, with a government subsidy, to 420 rupees, or about $6, per canister. That buys enough gas for a couple of months of regular cooking, although her sparing use stretches it to six. The price is a real barrier. "It is much easier to use the gas stove," Priti Kushwaha tells me. "It is just the inconvenience of getting it, and the cost."[7]

Fundamentally, the obstacle is the same one that keeps the air in big cities like Delhi so dirty: this country's grossly inadequate infrastructure. The distribution of LPG beyond cities is utterly insufficient to supply India's 650,000 villages with the fuel that could prevent so much suffering. So most days, Priti Kushwaha still burns wood and dung, which are not just free, but, in these parts, easy to gather. She uses gas only when she's really in a hurry. On top of the lantern rental business, she runs the sewing school her young neighbor Renu attends, so it's nice, when the days get hectic, to be able to turn a dial and watch the flame flare right up. Even her new biomass stove is far less burdensome than the traditional one, starting quickly, and burning more consistently, although she must chop wood into small pieces to fit inside. She has more time now for her children, and her other chores. I wonder what the poorer Kushwahas could do with that time if they had it: go to school more regularly, perhaps, or find work to supplement the family's income.

Outside Priti Kushwaha's kitchen, I gaze out over the village while Neha photographs a young boy and girl on a neighboring rooftop. Most of the homes look like they might topple over at any moment, but the people inside are resilient in the face of the hard truths that define their lives. Their struggle, I discover later, when I begin to understand the science of *chulha* burning, is not as far removed from urban India's troubles as it might appear. Cooking with biomass is now thought to be responsible for at least a quarter of India's overall air pollution.[8] With so many people burning smoky fuels so regularly, it makes perfect sense. So the nation's leaders—the people, far away in big cities, with the power to change life in impoverished villages like this one—may have more to gain from helping remedy their rural cousins' plight than they realize.

* * *

Although it takes me a while to realize it, my visit to Mustafabad has given me a glimpse of a shift in thinking among many of the scientists and policy experts trying to reduce household air pollution's toll. Efforts to address the scourge of dirty cooking fires have a bumpy history. For years, governments, aid groups, and international bodies promoted their "improved" biomass stoves, trying to give the poor a better way to burn wood, dung, and other smoky fuels.

Technically, it turned out to be a difficult task. What's more, while the latest ones are more user friendly, the stoves distributed in the past didn't deliver what impoverished cooks needed, and many ended up gathering dust. Until recently, the developing world's fires were seen more as an environmental problem than a health threat. So generations of stoves were designed not to safeguard human well-being, but to improve fuel efficiency in order to slow deforestation. Those doing the cooking found such devices decidedly inferior to what they already had. "Communities are very intelligent," one Delhi scientist remarked to me. "If I give you that stove, you will also not use it."[9]

Kirk Smith has helped change experts' approach, and he explains the shift to me when we meet, a world away from Mustafabad, in his office near the San Francisco Bay, on the edge of a hilly campus of towering redwoods and eucalyptus trees. "There was a lot of hype—drop some kind of gadget in the kitchen and it would fix it," recalls Smith, a global environmental health professor at the University of California, Berkeley. "There are people still at the World Bank who just laugh when you say the words 'improved stove.'"

Papers and cardboard boxes are piled everywhere in Smith's cramped workspace. He's rumpled and a little portly, with tousled gray hair, a bushy moustache, and glasses hanging from a string around his neck. For a long time, he tells me, scientists failed to grasp the severity of the danger from open cooking fires. It was a study he led in Guatemala that made it clear the old approach to reducing that risk was inadequate. Smith and his colleagues developed an air quality monitor babies could wear, and recruited subjects in the country's western highlands, where Mayan women often carry children on their backs as they cook. The infants were breathing in as much pollution as they would from smoking three to five cigarettes a day.

The research team installed brick stoves with chimneys in more than 200 homes, and found they reduced kitchen smoke by 90 percent. That seemed impressive, but the pollution measured by the babies' monitors fell by only half. Children, Smith realized, don't spend all their time in kitchens. They wander from room to room, and go outside, too, and doing so hadn't provided an escape from the fumes. "All the stove did was move the smoke, it didn't get rid of it," he says.

Like the team in Malawi, Smith's group tracked their subjects' pneumonia rates. And, as in Africa, the results were disappointing. It turned out the 50 percent drop in babies' smoke exposure brought only an 18 percent decline in incidence of the disease, although the number of severe cases fell more, by about a third. "Better than before," Smith says, but not good enough. Something more was needed.[10]

He was determined to find it. Propped on a shelf behind his desk is a photo of an Indian woman he first met in 1981, when she agreed to assist his research by wearing a monitor to test the air in her home while she cooked. They reconnected recently, when he visited her village. "It was a very emotional thing. She remembered me, she remembered the study." She doesn't cook anymore, but her daughter-in-law does; still, all these years later, over an open fire, a *chulha* stove. That was discouraging.

But something else happened on that trip, and it changed the way Kirk Smith thinks about household air pollution. One of the old woman's neighbors invited him into her kitchen. "She pointed at her LPG stove and said, 'I'm getting rid of that.'" He asked whether she was returning to open-flame cooking. Instead, she showed off an electromagnetic induction stove, a technology even cleaner than gas, one just beginning to catch on in wealthy countries like mine. Smith soon realized the electromagnetic stoves were on the shelves of stores in big cities like Delhi, and sales were booming. "That's when I said, 'I've got to start thinking out of the box.'"[11]

Today, he believes the best way to save the lives and the health of families like the Kushwahas is to give them access to the same cooking fuels we in developed countries—and wealthier urban Indians—take for granted. Rather than struggling to "make the available clean," as experts sought for so long to do with one new twist after another on biomass burning, he argues it is time to "make the clean available." That means gas and electricity, the only fuels good enough to truly safeguard their users. Of course, it's an insight Khushboo's big sister and cousins have already had for themselves; it was crystal clear to the young women I met that a gas stove would make their lives better.

I tell Smith about Priti Kushwaha's difficulties buying gas, and her sparing use of it. Such problems are common, he says, launching into a story about a speech he gave in India, urging oil and gas executives to invest in the infrastructure improvements needed to deliver LPG more widely. Smith didn't know it, but a top official from the government's petroleum ministry was in the audience. The two met later, for dinner. "He told me he went back to his office" after the talk and called in his staff to demand plans for expanding gas distribution to millions more households.

That kind of scale is exactly what's needed, Smith argues. "It's so refreshing, going to the Ministry of Petroleum. You say, '170 million [Indian households need a replacement for biomass],' and they say, 'OK, let's sit down and talk about it.'" Now, finally, India's leaders seem to have gotten serious about tackling the problem. In just a few years, working with fuel companies, officials have provided tens of millions of households with the equipment needed to start cooking with gas, and they've set ambitious goals for continuing that progress.

Of course, LPG is a fossil fuel, and fossil fuels drive climate change. The science of the soot—sometimes called "black carbon"—that comes from millions of wood and dung fires is more complicated. It doesn't linger in the atmosphere for centuries the way carbon dioxide does, but it can settle on glaciers or other ice sheets, where its dark color draws in heat and reduces the ice's natural ability to reflect sunshine, thereby accelerating warming.

So it's hard to say for sure what moving billions of people from biomass cooking to gas would mean for the climate. In Kirk Smith's view, it doesn't matter. Extinguishing all those smoky fires may well help the planet, but even if it doesn't, this is a question of justice. How, he asks, could we in developed nations, using fossil fuels for our cooking— and so much more to blame for the looming planetary crisis—deny those fuels to the world's poor, in the name of carbon footprints? It's not life-saving fuels for impoverished households that threaten the climate, he says. "It's you and me."[12]

* * *

As India starts on the immense task of bringing cleaner cooking to millions, it can look to the experience of nations further along the same path. Many are in Latin America, where reliance on biomass has fallen dramatically in recent decades, to about 15 percent of households.[13]

Brazil's relationship with cooking gas started long ago, by happenstance. It had been a stop on the *Hindenburg*'s route, and when the airship burst into flames while docking in New Jersey in 1937, ending dreams of travel by zeppelin, Brazil's leftover propane ended up in homes. Later, distribution networks grew so extensive they reached even remote settlements in the Amazon, where gas cylinders arrive by boat. Subsidies made the fuel affordable to everyone, and as LPG use climbed, wood and charcoal burning steadily fell. Those trends went into reverse after the subsidies were withdrawn in the first decade of the twenty-first century, and the cost of gas rose, forcing many to return to biomass. Economic growth, along with payments aimed more directly at the poor, have helped put things back on the right track.[14] Now, only 6 percent of Brazilian households use smoky fuels.[15]

Ecuador is going a step beyond that. It nearly eliminated biomass cooking with heavy LPG subsidies, and is now pushing its people toward electromagnetic induction stoves like the one that changed Kirk Smith's thinking. It's the newest—and cleanest—way to cook, fast and highly efficient. And Ecuador is expanding hydroelectric generation to power the stoves renewably. The government is providing them for free, or at a discount, to those who can't afford the purchase price, and others can pay in installments. Officials promote the new technology aggressively, on radio, TV, and social media.

Economic growth and urbanization help bring gas and electricity within reach, of course. But across Latin America, concerted government action is what succeeded in making them widely available. That kind of direct intervention, it's clear, is the fastest way to deliver change. "It's the vaccine analogy, it's the life-saving drug analogy," said one scientist I spoke to. "You don't wait for people to become rich before you get life-saving drugs to them. You step in and get them what they need."[16]

* * *

There's a surprising postscript to the story, one I encounter close to home. Because it's not just poor nations where wood burning takes a toll on health. In rich countries, too—particularly European ones—

the smoke wafting from fireplaces and log stoves adds a significant chunk to overall pollution levels. Of course, the scale is nowhere near what the developing world is grappling with, and it seems discordant even to place threats of such different magnitude side by side. But while the wood smoke drifting over Europe or North America is nothing compared with what Indians or Africans endure, when measured against a higher bar—that of healthy air—it is nonetheless a real problem.

For a long time, I had trouble taking wood smoke's dangers in the developed world seriously. I enjoy a roaring fire as much as anyone; it feels not just pleasantly cozy, but somehow natural, and thus surely harmless. So when I'd heard wood mentioned as part of the pollution picture in the places my reporting has taken me—the San Joaquin Valley, for example, or Krakow—I'd subconsciously discounted it, to focus instead on the threats that felt more pernicious: coal, diesel, agriculture. I understood, of course, that smoke as thick as the Kushwaha family breathes is deadly, but in countries like mine, it had always seemed merely rustic.

The truth is that wood smoke is thick with PM2.5 particles, and with toxins like benzene and formaldehyde. In Britain, I'm astonished to learn, it creates more than twice as much particulate matter as all the country's traffic (remember, diesel's threat comes largely via a different pollutant, nitrogen dioxide).[17] The gorgeous vistas around Mont Blanc in the French Alps are hidden in haze every winter as residents of the Arve Valley light up their stoves. Researchers in Helsinki reported wood burning accounts for as much as 29 percent of winter particle pollution there; in the suburbs, it reaches 66 percent. Greeks turned to cheap or scavenged wood for heat when economic crisis hit, and air quality plummeted. In California, heating with wood contributes more than 20 percent of wintertime PM2.5 emissions; at one site in Seattle, Washington, it accounted for nearly a third.[18] Wood smoke is why particle pollution in Fairbanks, Alaska, is among America's worst.

When particulate levels rise, of course, illness and death inevitably follow. Happily, the reverse is true too. A study that highlights wood smoke's unseen harms found death rates among elderly people in

Tasmania, Australia, fell by 10 percent after a push to convert households from wood to electric heat. And when more than 1,000 stoves were replaced in a small Montana town, particle pollution dropped by more than a quarter, and parents reported less wheezing and fewer respiratory infections among children.[19]

I'd just begun to understand all this when, early one January, I first noticed the smell of smoke on my block. Wood-burning stoves have become trendy in Britain, and I'm convinced a neighbor got one for Christmas. All winter, the smell wafted along our street every evening. Not unpleasant, unless I stopped to think about what was in it. And I soon realized how often I encountered that aroma; in fact, I knew exactly where to expect it, which of the blocks I routinely walk are home to someone kindling a daily fire.

Most people, I'm sure, simply have no idea how polluting those fires are. The worst offenders are not the stoves in which flames lick up behind glass doors, but ordinary, unfiltered fireplace fires. In fact, it's illegal to burn without a stove in most British cities and towns. But not only are the rules rarely enforced, most people don't even know they exist. And while the stoves are better than open fires, their contribution is still significant, and it grows with every new one installed.

Sometimes the consequences are easy to see. When an activist relays my request to hear from people affected, I'm shocked by the emails I get. From all around Britain, little villages and big cities, they tell of thick smoke from neighbors' chimneys, of futile struggles to keep it from seeping in through cracks around doors and windows, of laundry that reeks when it comes in off the line, of asthma attacks and lost afternoons in the garden.

Josephine Cock was one who wrote me. Her troubles began years ago, when she lived in Switzerland, where many of her neighbors lit fires every day. Although they followed all the rules—dried logs, approved stoves—the smell pervaded her home. She began to cough and wheeze; her chest hurt: "I felt truly terrible." She and her husband put up with it for years, but eventually they had to move. Now, the couple live in a small village on the Welsh coast; they chose it hoping the sea air would bring relief. But every year, more of their neighbors start burning wood. It makes Cock wheeze and her pulse race: "I go white in the face, my hands go blue, I shake, I feel sick and dizzy." She sends me a photo she took in a nearby town, when visiting her doctor: heavy smoke billows up from one corner, and wafts over everything. Near her home, one family's stove burns particularly regularly. She's sure they don't realize the harm they're inflicting. "They see the stove as something wonderful, warm, and cheap to run," Cock writes me. "They've come here, to the countryside, from an urban area and think they are truly 'back to nature.'"

Indeed, while many people light fires simply for the ambiance, others do so in the belief that burning wood is better for the climate than heating with natural gas. It's a view embraced by some environmentalists, trumpeted by stove sellers, and endorsed by governments that promote wood as "carbon neutral." The EU is a proponent, and many European countries offer grants and other incentives to encourage wood use. In Britain, subsidies come through a program called the Renewable Heat Incentive, and stoves carry labels proclaiming "eco" certification. Big wood-burning systems heat not just homes, but schools, recreation centers, government offices, and hospitals. Their operators boast they're doing their part to fight climate change.

Unfortunately, the truth behind those claims is more complicated. They depend, among other things, on new trees being planted to recapture the carbon released by wood burning. But that often doesn't happen, nor are the raft of other conditions on which the renewability hopes rest consistently met. If burning practices are less than perfect—wet logs, an overstuffed stove—efficiency dwindles and pollution increases. One scientist told me the carbon-neutral argument holds up only in "an ideal world which doesn't exist."[20] And just as with poor nations' cooking fires, the sooty black carbon in the smoke from European and American stoves makes its own contribution to climate change, potentially outweighing any carbon dioxide benefit.

Even where wood does deliver a climate win, the particulates released mean it comes at a cost. Burning in a city or suburb is especially dangerous, but particles travel long distances, so even fires in sparsely populated areas cause health problems. Few proponents of wood burning acknowledge the trade-off. And politicians, wary of being tarred as killjoys, rarely seem willing to step in, even to enforce existing rules.

The parallels to Europe's diesel disaster are troubling: a health-wrecking fuel sold on climate claims that may be mistaken, the embrace of such arguments by governments that ought to know better, their acceptance by ordinary people trying to do the right thing. The biggest difference is that while diesel's dangers are now, at last, widely recognized, wood burning has yet to attract similar opprobrium.

Indeed, it is growing, and threatening as it does to erase any progress cities make in reducing pollution from traffic. Not just heat, but electricity, too, now comes from wood. Old coal stations like Drax, Britain's biggest power plant, are being converted to biomass, often in the form of wood pellets shipped from the United States. On top of the health threat, an MIT study made clear the arguments used to justify such burning are flawed. At least in the near term, the researchers found, it's even worse for the climate than coal.[21]

It's disconcerting that just as the poorest nations struggle to break free of such smoky, health-draining fuels, we in wealthy ones are embracing them. Yes, wood fires are as old as humanity. But in the twenty-first century, on a planet as crowded as this one, we can surely—all of us—do better.

PART TWO

COMING UP FOR AIR

7

TO CHANGE A NATION

The Story of America's Clean Air Act

Across the middle of the country, smokestacks jut into the air, their gray-black plumes puffing upward. Sulfur dioxide, nitrogen oxides. Coal smoke. Heading high into the atmosphere, carried eastward. Changing, as it travels, into something new. Turning acidic. Then carried back to earth by rain, sleet, snow. Trickling into soil, stealing its nutrients. Freeing aluminum that kills roots, leaving trees to die of thirst and starvation, bare and brown. Poisoning lakes and streams. Suffocating fish whose gills are destroyed. Eroding buildings, statues, gravestones. The toxins that turn rain to acid seeping into human bodies too. But then, gradually, change comes. The power plants find a different kind of coal, low-sulfur. New scrubbers remove the ingredients of acid from their smoke. And far away, the air gets cleaner, the rain more pure. Forests and lakes, trees and fish, beginning to recover. The people healthier too. Progress. Incomplete, and imperfect. But real.

Tom Jorling and Leon Billings started as colleagues, but it didn't take them long to become friends. The 1960s were giving way to the '70s, and the two men were young Senate aides from opposing parties. The relationship they went on to build—like those among their more famous bosses—helped lay the foundation for a law that would prove to be one of America's most consequential.

During the workweek, Tom and Leon shared rides home in a pickup truck, and on Fridays they'd unwind over cheap beer and greasy sandwiches at their favorite local watering hole. Soon, their families were camping together in the Blue Ridge Mountains, on what started out as a tent platform and slowly grew into a house, Leon's retreat from Washington's stressful grind. There was a stretch when, by a quirk of geography and scheduling, Tom figures he celebrated more birthdays with the Billings family than with his own wife and daughters. At Leon's memorial service, almost a half century after they first met, he would describe the corner of Virginia they used to visit as the closest thing to a piece of Montana his friend could find near his adult home.

Montana was where Leon grew up, the son of radicals on the edge of the Great Plains. His parents' dinner table was a regular stop for all manner of progressive leaders, the tab to be paid with good conversation and a bottle of whiskey. Harry and Gretchen Billings ran a crusading, union-backed newspaper, and in a state dominated by copper mining interests, its agenda didn't make their three sons popular. Leon sold copies of the *People's Voice* to state workers in the capitol building, some of whom joked that his red hair was the same hue as his parents' politics. If he could hold up his end of the sharp banter, he learned, even an ideological opponent would buy a paper. "I really cut my teeth on shooting my mouth off," he'd recall, "something I never forgot how to do."[1]

Later, when Leon Billings built his own career in government, the word "abrasive" was often used to describe him. He was loud, foulmouthed, and certain in most any disagreement that he was in the right. "There was only room for one opinion," a former underling recalls.[2] His friend saw past the tough exterior, though, to the generous soul underneath. Tom was from an altogether different background, a football star at his Jesuit high school in Cincinnati who enrolled in officer training at Notre Dame, and seemed headed for Vietnam until an old knee injury disqualified him. Although he was in his twenties in the drugged-out 1960s, Tom didn't get his first whiff of marijuana until he took a job in the U.S. Senate, where antiwar protesters camping out in the corridors smoked the stuff all night.

That was where he and Leon met in 1968, staffers on a still-obscure subcommittee. Tom didn't have a strong party identity, but he was the lawyer for the panel's Republicans, while Leon was counsel to the Democrats, and the two could have become adversaries. All these years later, his gait a bit halting now, his eyebrows bushy, Tom Jorling has borrowed an office in Albany, New York, for our meeting. It's been only a few months since Leon died suddenly, of a stroke, exactly a week after an election that would bring a grave threat to bear against his life's proudest achievement. So while Tom is warm and talkative, full of stories, he grows quiet every now and then, gazing into the distance, and it's clear the grief at his friend's death is still fresh.

The connection he and Leon forged when they were 28 and 30, he tells me, still feels a little mysterious. "We developed a kind of simpatico thinking process. He'd say something, and I'd say something, and he'd say something. It all fit together." Times were turbulent, but the bitterest divisions back then—over civil rights for black Americans, and the war in Vietnam—fell not along party lines, but regional ones.

Leon reported to Senator Edmund Muskie, the lanky former governor of Maine, known for his towering intellect and a temper even more volcanic than his aide's. He'd been the Democrats' vice presidential nominee in 1968, and despite the defeat he and Hubert Humphrey had suffered, Muskie was widely seen as the favorite for the top of the ticket in '72, when President Richard Nixon would be up for reelection. Tom's boss was Senator John Sherman Cooper, a soft-spoken Kentucky Republican with a strong independent streak; he had been among the first in his party to stand up against McCarthyism and later became a champion of civil rights and leading antiwar voice.

Environment was a word that had barely entered Americans' political vocabulary when Tom and Leon met. But pollution was front-page news, and the public seemed to want politicians to do something about it. When Cleveland's Cuyahoga River caught fire in 1969—it was actually oil-soaked debris burning on the surface—it crystallized a feeling that the time had come to reckon with the fallout of industrialization. In cities across America, the air was no cleaner than the rivers. Not just famously smoggy Los Angeles and steel-making Pittsburgh: even in Washington, there was a coal-burning plant a few blocks from the Capitol, and Hill staffers had to change their soot-stained shirts by midday. In New York, ash from incinerated garbage floated through the sky, and, at the bottom of Central Park Lake, scientists later discovered layers of the lead that had permeated the air.[3] "On the autopsy table it's unmistakable," a coroner told the *New York Times*. "The person who spent his life in the Adirondacks has nice pink lungs. The city dweller's are black as coal."[4] It was a story repeated across the country.

When Ed Muskie first came to the Senate, he'd hoped for a seat on a high-profile committee like Foreign Relations. Lyndon Johnson was the brass-knuckled majority leader then, though, and Muskie had crossed him, so he found himself assigned to the backwater of Public Works instead.[5] He'd seen Maine's rivers poisoned by paper mills, so he asked to set up an Air and Water Pollution subcommittee, and he became its chairman, staking out jurisdiction no one else had claimed. His panel convened hearings on the awful air, taking testimony in cities all over America. In those days, Tom Jorling tells me, pulling a stack of index cards from his shirt pocket to demonstrate, the main way to gauge pollution levels was to hold up a color-coded card and eyeball it to judge which panel best matched the sky. Muskie's hearings set out to determine why previous attempts to address the problem, a series of clean air laws in the 1950s and '60s, had failed.

Many of the subcommittee's members would go on to etch their own names in Washington's history. Democrat Thomas Eagleton would briefly be his party's vice presidential nominee, until the news he'd undergone electroshock treatment for depression got him pushed off the ticket. Bob Dole, on the Republican side, would make his own run for the Oval Office. Howard Baker, like Dole a future Senate majority leader, actually made it to the White House, as Ronald Reagan's chief of staff. His most enduring fame would come from the piercing questions he asked—about the leader of his own party—at the Watergate hearings: "What did the president know, and when did he know it?"

That was all in the future, though, and despite the outsized ambitions on the subcommittee—and the members' awareness that Muskie, their chairman, would likely soon be his party's presidential nominee—the men worked together with remarkable comity. Back then, hearings were for gathering information, not reaffirming existing ideologies or showboating (there were no cameras around to preen for), and the senators routinely stayed till the end, listening and asking questions. It had become clear that progress on air pollution would require a new law, so their work shifted toward drafting it. What tools would the government need to deliver cleaner air, the senators asked themselves, and what powers did the Constitution allow? "It was a very high-quality intellectual engagement, with a commitment to produce a product that would deliver the result," Tom Jorling tells me.

Party hardly seemed to play a role. The senators "each had individual views on certain things, but you couldn't tie that individual view back to a partisanship," Jorling says, running a hand over his white hair as a golden retriever pads around outside the sunny corner office where we're sitting. "They had a tremendous amount of respect for each other, they listened to each other, they enjoyed each other." Muskie was known for his puns—some better than others—and Baker and Eagleton also used wit, bawdy at times, to break the tension of long hearings. The two sides trusted each other too. Democrats would sometimes call Tom, the Republican counsel, with questions, and the Republicans would similarly phone Leon. Muskie and Baker's friendship was particularly warm. Nearly unimaginable from today's vantage point, such cooperation was unusual even then. "'Why?' is the question that Leon and I continued to ask each other," Tom tells me. One answer, they concluded, was simply the quality of the individuals. But the aides put their finger on something else, too, a singular experience the senators had shared: All but one of the subcommittee members had served in World War II. "They went through something in a way that said, 'Democrat, Republican, it doesn't make any difference.'"

Muskie had an important insight about why earlier efforts on clean air had proved ineffective: They had let questions of cost and technical feasibility shape rule making. But only industry could judge what was economically and technologically possible, he realized, putting regulators at a perennial disadvantage. The new law's most radical achievement would be to demote such considerations to second place. From now on, public health would be the sole factor used to determine how much pollution was allowed in the nation's air. Cost could come into play later, in deciding how to meet the new national standards, but only physical well-being would be taken into account in setting them. That reframing—the decision to put Americans' health above companies' profits—was the beating heart of the bill, the fundamental reordering of priorities that would invest this new Clean Air Act with a power its predecessors had lacked. Not surprisingly, it was also the decision that would, before long, make the law a target for opponents pushing relentlessly to undermine it.

The system the subcommittee created gave each state the power to design measures suited to its own circumstances. Washington would set air quality standards, with limits on each of a list of pollutants. And every state would have to submit a plan for meeting them. Once approved, those plans became legally binding. There were more pieces to the puzzle too. A law, Tom Jorling says, is only as strong as its weakest link. So he and Leon, and their bosses, the senators, had to make it fail-safe, so the job couldn't be left unfinished. He's brought a canvas bag stuffed with papers to our meeting, and as he explains each senator's contributions, Jorling rifles through a tattered copy of the bill's official legislative history. The binding of the thick green paperback is broken, several big chunks are falling out, and colored sticky notes poke from many of its pages.

Senator Eagleton, he recalls, said Americans must be able to trust the law would fulfill its promises, so he proposed deadlines for achieving the air quality standards. They'd need to know it would be enforced, so the bill would be the first to include a new kind of provision, known as "citizen suit," empowering people to take their government to court if it didn't do its job. One wording choice was especially critical: In setting out the government's power to act, the bill's authors used the mandatory "shall" instead of the weaker "may," so even reluctant administrations would have to uphold it. The Clean Air Act would bring an unprecedented level of accountability to the protection of Americans' health.

Senator Baker was a believer in innovation, and he wanted to push industry to develop new ways of doing things, not just use technology it already had. If government laid out tough requirements, he felt, engineers would respond, and figure out how to achieve them. So the senators set a high bar for automakers: In five years, every new car would have to be 90 percent cleaner. Howard Baker's idea became known as "technology-forcing regulation," and although Congress would later extend the timeline, he was proved right. Not only did car companies meet the senators' ambitious goal, they achieved even tougher standards later. And the new technologies they developed—foremost among them was a device called the catalytic converter—would save countless lives, not just in America, but around the world, as the new equipment gradually became standard nearly everywhere.

The senators could see how quickly scientists' understanding of air pollution was advancing, and they didn't want their bill to become obsolete. So they required regulators to review the limits every five years, and update them in accordance with the latest science; they could add new pollutants to the list as well as changing the permissible levels. The law also said its provisions could apply to pollutants not yet identified, a future-proofing that would have important consequences decades later.

It was up to the two young staffers to turn their bosses' ideas into legislative language. Each evening, they lugged their dictation machines into Leon's pickup, and spent the ride home drafting sections of the bill and brainstorming strategy. Once, Tom's wife has reminded him, "we were so engaged, we got out of the car at our house, went in, and—she had already set the table for dinner—cleared off the table and put down the committee prints and started working on them." Those were the clerk's cleaned-up versions of Tom and Leon's drafts; the time needed to print each draft allowed a deliberative pace lost in today's technology-driven frenzy.

Of course, the companies on whom the bill would force change fought bitterly, predicting financial ruin. Tom says they got a fair hearing, although a story Leon's friends like to tell makes his views clear: At one meeting with industry representatives, he folded the papers they'd handed him into airplanes and floated them across the room. There was no mystery about where that antipathy had come from. It was Leon's boyhood that drove him, the example of parents who spent their lives trying to make the world more just. From today's vantage point, the notion that we all have the right to breathe healthy air hardly seems radical. But the law Leon Billings was helping to write—in its assertion that the health of human beings was more important than the profits of corporations, and that their government had both the power and the responsibility to protect them—was nothing short of revolutionary. That was what drew him to the cause, his fight in some ways an extension of the lonely one his mom and dad had waged in Montana. He and those standing beside him "were out to change the world," to make big companies do what was right, one friend told me, "and this was their tool." For Leon, there was no other choice. "It was in his blood."[6]

The clock was ticking, since any bill that was not ready for the president's signature by year's end would die, months of effort wasted. Meanwhile, Americans were waking up to the issue's urgency. Millions turned out for the first Earth Day in April 1970, and a Nixon aide described public opinion on the environment as a tidal wave "about to engulf us."[7] The House of Representatives passed a bill weaker than the Senate's, but little of its version survived; it was the work of Ed Muskie's subcommittee that would last. In the end, the bill won unanimous Senate approval and drew only one nay in the House. It was technically a set of amendments to an earlier law, but the legislation that emerged from Congress that December would become known simply as the Clean Air Act of 1970.

Richard Nixon didn't harbor much enthusiasm for the fight against pollution. Not long before he died, an aide told him he'd be remembered as a great environmental president. "God, I hope not," he was said to have replied.[8] But Nixon had a sharp political sense, and he saw claiming this issue for himself would deny Ed Muskie, his likely opponent, a cudgel to use against him in what was looking like a tough reelection campaign. So President Nixon created the Environmental Protection Agency. And at noon on the final day of the year, he signed the Clean Air Act into law, a landmark moment for public health, and the start of a new era in America's modern history.

* * *

The threat Nixon had sensed from Muskie never materialized. The senator fell victim to one of the "dirty tricks" that later mushroomed into Watergate. Just before the New Hampshire primary, reporters claimed they'd spotted a tear on his cheek as he responded to smears against him and his wife (later revealed to have come in part from the president's camp). He denied having cried, but even the hint of it sank his campaign. George McGovern, a much weaker candidate, got the nomination instead, then found himself on the wrong end of a landslide in the fall.

But the law that would be Edmund Muskie's lasting legacy soon began to make its power felt. Around the country, states erected the regulatory architecture they'd need to achieve the targets Washington set. There is no silver bullet for dirty air, no single action that dissipates it. Cleaning up requires a systematic effort, sustained over time, to identify pollution's sources and require action from those responsible. As the 1970s wore on, that's exactly what began to happen. At the behest of the new regulators, thousands of changes, big and small, took root. Factories and power plants installed scrubbers and other pollution controls, cities closed garbage incinerators, states introduced auto inspections and cracked down on the vapors wafting from filling stations. As carmakers grappled with the challenge Howard Baker had given them, the EPA required unleaded gasoline be sold and mandated lower-sulfur diesel.

Not every deadline would be met, but together, the standards pushed polluters in the right direction. And slowly but surely, the air began to clear. Twenty years after the Clean Air Act's passage, carbon monoxide levels were half what they would have been without it, particle pollution was 45 percent lower, and sulfur dioxide and nitrogen dioxide were down by 40 percent and 30 percent, respectively, a government analysis found. In just one year, 1990, the law saved about 200,000 lives and averted 150,000 hospitalizations for heart and lung trouble. Thanks to a 99 percent drop in airborne lead—a wrecker of brains—American children were collectively spared the loss of 10 million IQ points in that single year. Perhaps even more astonishing were the economic effects of all that improved health. The study put the law's benefits at $22 trillion in its first two decades: 44 times the cost.[9] The Clean Air Act of 1970 was far from perfect, and there was more work to do. But about one thing, there could be no doubt: This law had changed America for the better.

* * *

It was also making enemies. So the Clean Air Act would need defenders, and Henry Waxman was one of the fiercest. Waxman is not an eloquent orator. He's not especially telegenic, nor famous beyond the world of politics. What he is, though, is tenacious. "Don't get into an argument with Henry," a colleague once warned. "But if you do, bring your lunch."[10] It was that doggedness, and a talent for finding, and pulling, every lever available to him, that made Waxman one of modern Washington's most effective lawmakers. "Probably no living member of Congress has accomplished as much as he has," one writer pronounced when Waxman left office in 2014. "I'm not sure many dead ones have, either."[11]

Waxman was 35 when he won his House seat in 1974, along with a wave of other newcomers, mainly Democrats like him, elected by voters looking for fresh leadership after the agonies of Watergate. He brought with him a deep-seated belief in the power of government to be a force for good, a protector of ordinary people. Lou Waxman, a grocer, always told his son how President Roosevelt had helped families like theirs weather the dark years of the Great Depression. "Business only looked out for its own," Henry Waxman later wrote, summing up his dad's view. Government's power "ensured that the little guy had a chance."[12]

Another thing he brought to Washington from Los Angeles was an insight he'd picked up as a California state lawmaker: Developing expertise on one issue could give a new arrival more sway than he was due in a body governed by seniority. Waxman chose health, and it was that focus that drew him to take up Tom Jorling and Leon Billings's cause and set him on what would become a career-long commitment to not just protecting, but also building on, the law they'd helped write.

His work on the Clean Air Act was far from his only passion. Waxman would also push through several expansions of health coverage, food and water safety protections, and regulation of tobacco companies, among dozens of other accomplishments. "Fifty percent of the social safety net was created by Henry Waxman when no one was looking," an observer once quipped.[13] When he couldn't scrape together enough support to pass legislation, he'd hold hearings to draw attention to it. When Democrats were in the minority and he couldn't convene hearings, he'd write reams of letters, figuring a congressman's stationery could at least open some doors. "'Bulldog' Waxman," one headline called him.[14] "Liberalism's legislative genius," proclaimed another.[15]

For a relentless bulldog, Henry Waxman is pretty personable. He's late for our meeting at the consulting firm his son runs, and comes in apologizing. He's slight, with a shiny bald pate and salt-and-pepper moustache, an open-necked shirt under his navy blazer. He's clearly mellowed since retiring from politics. "I eat lunch every day now that I'm in the private sector," he laughs.

In 1977, not long after Waxman arrived in Washington, Congress tweaked the Clean Air Act with a set of amendments. But the atmosphere in the capital would soon change, with Ronald Reagan's election at the start of the new decade. The popular new president's political philosophy differed profoundly from Waxman's. Government, in Reagan's view, was an obstacle that stood in the way of American progress, not a tool for making progress possible. "The nine most terrifying words in the English language," Reagan famously said, were "I'm from the government, and I'm here to help." That belief, growing ever more fervent, would stand as his party's bedrock in the years to come.

The Democrats controlled the House, and the Republican president had an important ally among them: John Dingell. The Energy and Commerce Committee chairman was sometimes known as "the Congressman from General Motors." Dingell's Michigan district was home to all three big automakers; his wife worked at GM for decades, and the couple lived a mile from Ford headquarters. So it made sense that he was the industry's most reliable champion. He and Waxman had a complicated relationship. On many issues, they were on the same side. But on anything that affected cars, Dingell was a powerful foe.

Dingell was the point man for Reagan's push to roll back the Clean Air Act's protections. They had the strong support not just of carmakers, but the coal, oil, steel, power, and chemical industries, for whom the law was an expensive headache. Waxman marshaled his forces as best he could. "When confronted by a steamroller, as we were about to be, you first need to slow its momentum," he wrote years later. "One way is to stall," with parliamentary maneuvering and a blizzard of amendments. "The other way is to win a skirmish."[16]

Waxman did both. He used every tactic he could think of to drag out debate, hoping that, with midterm elections looming, delay would magnify the power of his most potent weapon—public opinion. A 1981 poll had shown 86 percent of Americans opposed weakening the Clean Air Act, and Waxman invited the pollster to testify. Lou Harris's words were chilling for Reagan and Dingell's forces: "Never in my career have I seen such strong opinion on one side of an issue,"[17] he said. Try to undermine this law, he warned, "and you will see that you lose the '82 election."[18]

Nonetheless, Dingell's committee voted down almost 60 of Waxman's amendments, and things were looking bleak. Hoping to turn the industries he was battling against one another, Waxman targeted just one, the chemical makers, with an amendment on their plants' carcinogenic emissions. After it won by one vote, those companies pulled their support, and the alliances Dingell had built collapsed.[19] The effort to kneecap the Clean Air Act had failed.

But saving the act wasn't enough. Now Henry Waxman wanted to update it. A new challenge had come to light, one the original law had not foreseen: Acid rain was plaguing the Northeast and parts of Canada, killing trees and poisoning lakes. The culprits were midwestern power plants burning high-sulfur Appalachian coal—hazardous to Americans' bodies as well as the nation's landscape—and to stop it, they'd have to install expensive smokestack scrubbers. The industry was resisting, and things seemed to have reached a stalemate. Unable to strengthen the act, Henry Waxman would spend the rest of the Reagan years working toward the day when he could.

Eventually, he spotted opportunity in a faraway crisis, the toxic leak that killed thousands of people near a Union Carbide pesticide plant in Bhopal, India. The original Clean Air Act had failed to adequately address toxic chemicals in the air, and doing so was one of Waxman's top priorities. So he brought his subcommittee to West Virginia and held hearings beside another Union Carbide plant, one producing the same gas, methyl isocyanate, that had devastated Bhopal.[20] As he'd hoped, the hearing drew public attention, and he pressed his advantage with a plan to toughen airborne toxin rules. By the time the bill made it through Congress, the only piece that survived was a provision for conducting a national inventory of toxic chemicals. It was a small victory, but Waxman thought it might pave the way for a bigger one.

He was right. By the time the first toxic inventory report was published in 1989, the political sands had shifted. George H. W. Bush was in the White House, having promised to be "the environmental president." And with hypodermic needles washing up on the Jersey Shore and oil gushing from the wrecked tanker *Exxon Valdez* in Alaska, such concerns were in the news again.

Waxman didn't much like the clean air bill Bush proposed, but he could work with it. One of his biggest objections was a change to a word the 1970 bill's drafters had chosen carefully: The White House version said the EPA "may"—not "shall"—enforce pollution standards. Like Tom Jorling and Leon Billings, Henry Waxman understood the difference. And like them, he was determined the stronger language would prevail so future administrations wouldn't have a choice about applying the rules. On that crucial question, the bulldog got his way.

Acid rain was not the only problem that had come to light since 1970. A dangerous hole had opened up in the ozone layer, far above Earth, letting through skin cancer–inducing rays. The cause was a class of chemicals called chlorofluorocarbons, used in refrigeration and as propellants in spray cans. And Waxman also wanted to toughen rules on the smog still blighting many cities.

This time, Dingell's hand was weaker, and rather than fight, he chose to negotiate. "It's a process with which I am not entirely comfortable," the midwesterner said wryly when a deal was announced. "But it has been a success."[21] The bill that came out of Congress, with overwhelming bipartisan support, was far stronger than the one that had arrived on its doorstep.[22] It created a market-based "cap-and-trade" system to reduce emissions of the sulfur dioxide creating acid rain (and sickening those who breathed it), required the phaseout of the chemicals causing the ozone hole, and gave the EPA a list of 189 carcinogens and other toxins to regulate. It also strengthened rules for areas failing to meet the original air quality standards. It had taken a decade, but Henry Waxman got his victory. Bush signed the Clean Air Act Amendments of 1990 in November, the second Republican president to approve a life-saving air quality law.

The bill would dramatically reduce acid rain, and would allow the United States to play its part in the international effort that helped the ozone hole begin to close. And it was another big step forward for the nation's health, ensuring the air Americans breathed would continue to get cleaner. By 2010, the measures Henry Waxman had pushed through were saving 160,000 lives every single year, and the government predicted that annual figure would rise to 230,000 by 2020. Economically, the benefits have been 30 times greater than the costs,[23] an extraordinary refutation of industry's dire predictions.

In Henry Waxman's view, the amendments' passage proved a few other things too. Perhaps most importantly, he believes, the win vindicated his lifelong belief that congressional power, when used wisely, can change Americans' lives for the better.

Of course, the fight was never really over. Waxman would have to push to make sure future presidents—including Bush's son—enforced the law. Once, a reporter told him the 1990 amendments looked inevitable in retrospect. Thinking back on the years of battles, the powerful foes working to weaken the bill, all the ways things could have turned out differently, he knows that wasn't so. "We fought so hard, over so long," Henry Waxman says. "It's never-ending. But that's part of the job."

*   *   *

Christine Todd Whitman played a part in the Clean Air Act's long story too. When she picks up the phone at her farmhouse in New Jersey's rural western corner, I immediately remember the combination of warmth and confident straightforwardness that made her a successful politician. I'd already left the state when she became its first (and only) female governor, but across the river in New York, I'd often see her on the news. Later on, she became responsible for enforcing the Clean Air Act, as head of the EPA under President George W. Bush.

It was an awkward fit from the start: Whitman was a patrician East Coast Republican among Texans in cowboy boots, a moderate among conservatives, a former governor proud of her work protecting New Jersey's coastline in an administration dominated by oilmen. She would have preferred a different job, maybe UN ambassador, but that post included responsibility for speaking on reproductive issues, and her vocal support for abortion rights was out of step with Bush's staunch opposition. So she ended up at the EPA, where it didn't take long for the cracks to start showing.

Her adversary was the man many saw as the power behind Bush's throne: Vice President Dick Cheney, whose many responsibilities included heading the White House's Energy Task Force. In the administration's first months, before 9/11 upended its agenda, a new national energy policy was a top White House priority. California was suffering through rolling blackouts and electricity price spikes, and Cheney's team was crystal clear on why. While more neutral observers attributed the crisis to deregulation and market manipulation, the task force blamed the Clean Air Act. "I said to them, 'Show me permits that are being held up'" by air quality rules, Whitman recalls. "They never came up with one, because it was economics, not environmental regulation" that was the real culprit.

Although Bush had renounced the Kyoto climate change agreement, Whitman felt she had a meaningful peace offering to bring to a meeting on global warming with environment ministers from America's G8 allies in her first weeks on the job, one she hoped would ease their worries about a president many feared would run roughshod over environmental concerns. George Bush, Whitman knew, had promised during his campaign to limit carbon emissions from American power plants. The pledge was reiterated in the transition briefing books she'd been given. So before the trip, she told Condoleezza Rice, then national security adviser, she wanted to highlight it at the meeting, and Rice approved. So, Whitman later wrote, did the White House chief of staff's office.[24]

The promise went down well at the Trieste summit, where Whitman's words were taken as a sign the Bush administration would treat climate change more seriously than many had imagined. But even before she boarded her flight home, opponents were mobilizing, and a group of Republican senators wrote to the president, arguing against the carbon cap. Shortly after her return, Whitman was called to the White House, where Bush told her he was retracting his promise. He apologized for leaving her out on a limb, but said the looming energy crisis prevented him from imposing a new burden on utilities. As she left the Oval Office, she recalled in her memoir, she passed Vice President Cheney in the narrow hallway right outside. He was in his overcoat, clearly in a hurry, and someone handed him a letter he tucked into his pocket. Whitman would later learn it was Bush's reply to the senators who'd urged him to drop the carbon limit.[25] He not only did so, but also went a step further, saying he did not believe carbon dioxide was a pollutant covered by the Clean Air Act.[26]

George Bush did not get the last word on that question, which would eventually make its way to the Supreme Court. But in the meantime, Christie Whitman hurried to reach her counterparts in London, Ottawa, and the other G8 capitals. "I had to call them and say, 'You know what I told you last week? It's not gonna happen,'" she tells me. "It was frustrating and embarrassing." The decision "kind of flipped the bird to the rest of the world, who cared a lot about carbon." In that sense, it was of a piece, Whitman believes, with the go-it-(nearly)-alone attitude that so damaged America's alliances on the way to war in Iraq.

In this case, it wasn't just the country's credibility that was damaged, but Christie Whitman's, too, and that made an already difficult job even harder. "This was about, 'Do I really speak for the administration? Do I really know what's going on?'" Such turnabouts grew familiar, and she'd sometimes commiserate with another administration moderate, Secretary of State Colin Powell. "We each had the same experience," she recalls, of meeting with an agreeable president, then getting a call saying the White House position had changed.

What stunned Whitman most was the vehemence of her fellow Republicans' anti-regulatory feeling. She'd expected opposition from Democrats and environmental advocates. And she shares her party's small-government principles and desire to reduce red tape. "I get it," she tells me. "It's easy to hate regulation," because it costs money. But Christie Whitman also knows the other side of the story, understands environmental rules have brought enormous strides in protecting Americans' health. "Regulations are based on scientific research, there's a reason behind them," she says. "It wasn't that somebody at EPA got up in the morning and said, 'OK, whose life can we mess up next?'"

Republicans weren't the only ones giving her headaches, of course. Environmental organizations, in her view, are often too ready to hype every problem into a crisis, a tendency she ascribes to their need to keep supporters inflamed enough to give money. And to be sure, Whitman has her own problems with the Clean Air Act. The rigor its framers built in sometimes felt to her like inflexibility. And exhausting litigation—from advocates demanding tougher implementation and industries petitioning for a looser approach—has marked every step of the act's history. "Anytime you did anything with it, you ended up in court," she says. Yet Whitman knows the Clean Air Act's successes coincided with a decades-long expansion of the American economy, so the right's angry claims that regulation stifles growth seem spurious to her. We use far more electricity, drive many more miles, than we did in 1970, she points out, but because government has told companies they had to build better power plants and make cleaner cars, the air we breathe is far healthier.

That wasn't how Dick Cheney saw it. The breaking point came in 2003, over an obscure-sounding Clean Air Act provision: If a utility undertook extensive enough renovations on an old power plant, the act said, it would be subject to the tighter pollution rules that governed new plants. Routine maintenance wouldn't trigger that expensive reclassification, but no one had ever defined what routine meant. Many were honest, Whitman says, but some utilities would pass off major renovations as minor. Her staff had been drafting a proposal on where to draw the line. But Cheney had his own ideas, and he'd been pushing her to move quickly to ease the rules. Finally, "they handed me a bunch of numbers" that she could see would give cheating companies a yawning loophole, she tells me, and she suspected they'd come straight from the energy industry. This was a big deal for the White House, and Bush's chief of staff finally said, "Enough of this. This is the standard, and go do it." Instead, Whitman quit. "They had a right to have an administrator who would sign it in good conscience and then implement it. And I just couldn't do that."

The public explanation was that she and her husband were tired of a commuter marriage. They did hate it, but she's clear now that the dispute with Cheney was the real reason for her departure. In truth, anyone could see she'd been a misfit in the administration. "The EPA's lonely moderate," a *Washington Post* headline called her. Indeed, "moderate Republican," the label with which Whitman is most often tagged, was already becoming a lonely one to wear in 2003. Tom DeLay, the Texan known as "the Hammer," had risen through the House leadership after branding the EPA "the Gestapo of government." He was far from the only Republican with such views.

*   *   *

The Obama years would mark another turn in the Clean Air Act's history. In 2007, with a decision known as Massachusetts vs. EPA, the Supreme Court had paved the way for a new application of the old law. The justices said the act empowered the agency to regulate carbon dioxide, which doesn't harm the health of those who breathe it, but is the chief cause of the planet's dangerous warming. Indeed, it said the EPA must regulate carbon unless it could provide a scientific basis for a decision not to. "Greenhouse gases fit well within the Clean Air Act's capacious definition of air pollutant," Justice John Paul Stevens wrote.[27]

Obama would have preferred a law explicitly aimed at the climate threat. But when the last hope of congressional action on the issue died in 2010, with the demise of a bill bearing Henry Waxman's name, the president turned to the 1970 act. The foresight of Edmund Muskie and his colleagues, their awareness of what they didn't yet know, gave

Obama the power to undertake America's most sweeping effort to reduce its greenhouse gas emissions. He set out new rules on power plants and ambitious auto mileage standards, putting the United States in a position to promise the world its carbon footprint would shrink. That promise helped make possible the Paris Agreement, the most concerted global effort yet to stave off runaway climate change.

Not surprisingly, Obama's actions were contentious. Republicans directed their fury at the agency charged with carrying out his plans: the EPA, or, as Senator Marco Rubio branded it, "the Employment Prevention Agency." Fueled by oil and gas money from donors like Charles and David Koch, the rhetoric grew ever more incendiary. Seeking to undermine the justification for Obama's rules, the House of Representatives science committee subpoenaed raw data from 20-year-old Harvard and American Cancer Society studies demonstrating air pollution's harms. Senator James Inhofe said his granddaughter was among the American children "brainwashed" about climate change and brought a snowball to the Senate floor to demonstrate such worries were overblown. And of course, there was the Twitter-happy real estate mogul who, with an eye on the next election, branded climate change a hoax, a Chinese invention aimed at wrecking America's economy.

Congressman Pete Olson isn't that sort of fire-breather. He's charming in a slightly goofy way, and he banters about sports as he leads me into his office off one of the grand marble hallways of the Rayburn House Office Building. Olson, a Republican, represents Houston and its suburbs, and hanging near his receptionist's desk is a black-and-white photo of an old oil well spewing a spray of thick smoke. Houston, of course, is oil country, and the energy industry is Olson's biggest financial supporter.[28]

When he moved to Texas as a young boy, Olson tells me, the smog was so thick the downtown skyline was visible only 10 days a year. "Now, that's flipped," he says. "I can see downtown every day except for maybe 10." That, he knows, is thanks to the Clean Air Act, and he readily acknowledges the good it's done.

It's the way the act has been applied more recently that bothers him. As air quality has improved in places like Houston, he says, the EPA has continually shifted the goalposts, burdening local officials with endless paperwork and penalties, and bringing areas whose air was once deemed good enough into noncompliance. "They're throwing their hands up and saying, 'We can't achieve that,'" he tells me. What's more, Olson continues, federal regulators don't cut regions any slack when droughts or forest fires make standards harder to reach, nor do they account for the pollution drifting from beyond America's borders, or the high natural ozone levels in some places.

So the congressman has been pushing a bill that would loosen the Clean Air Act's requirements. He says he wants to keep what's worked in the law, but strike a better balance between public health, practicality, and cost. Among other things, his proposal would allow the EPA to consider feasibility alongside health in setting air quality standards. That strikes at the act's very foundation, in part by handing local officials uninterested in tackling pollution an easy excuse for failing to achieve federal standards. So the act's defenders see warm words like Olson's as deceptive cover for an attack that would make America's air dirtier. It's easy, and popular, for politicians to say they want clean air. But it makes no sense to will that end without supporting the means of achieving it.

There is, of course, a philosophical grounding to measures like Olson's, and it's dramatically different from the one that guided Tom Jorling and Leon Billings. A few blocks from Capitol Hill, Diane Katz is all curly dark hair and chunky glasses as she greets me in the Heritage Foundation's lobby. Katz hasn't worked on Olson's bill, but as the influential conservative think tank's regulatory expert, she often advises congressional Republicans and staffers at federal agencies— the people with the power to write, or repeal, rules that change lives. Fox News is on the big TVs in the cozy lounge where she makes us coffee, and as we talk, the foundation's president appears on screen, being interviewed from somewhere in this building. It's a pointed reminder of the sway Heritage and its views hold in Republican Washington.

Like many conservatives, Katz tells me, she started out on the left. As a young social worker in a halfway house for recently released prisoners in Ohio, she noticed some failed to seize opportunities to better themselves, and concluded that "government was pretty limited in its power to change human behavior. Very limited, actually. And it led me to start questioning whether we're expecting way too much of the government." She moved to Washington, where her ideological conversion was sealed when, as a journalist, she saw its workings up close and became convinced officials' good intentions often had damaging consequences. Katz acknowledges the progress the Clean Air Act has delivered, but she sees it as a case study in federal heavy-handedness. She doesn't know whether that is in the nature of the law, or the people enforcing it, "or whether it's inexorable—you give an agency that much power and ultimately it's going to end up using it, and more." When dramatic air quality improvements left regulators with little to do, Katz argues, they grasped for a justification to push further. "That's what led to the emphasis on climate change, because it was almost as if they were looking for a new cause to coalesce around," she says. "That's where I think things really started going downhill."

Diane Katz's take is the twenty-first-century update of the philosophy Ronald Reagan ensconced in Washington, the view of government as an overweening bureaucracy more likely to make trouble than solve it. For one of the nation's two major political parties, the EPA—flawed, to be sure, but doing its best to protect Americans' health, carrying out the mission a Republican president entrusted to it—now stood as a dangerous enemy, an out-of-control job killer needing nothing more than to be cut down to size.

* * *

Tom Jorling had promised his family a life in the countryside, so in 1972 he quit his Senate job to teach at Williams College in Massachusetts. Before he left, he and Leon Billings followed up their Clean Air Act work with another achievement, helping Senator Muskie's subcommittee write the Clean Water Act. These were the foundational laws of American environmental protection, and in the years before Leon's death, the pair taught a course on the history they'd been part of. They chewed over, too, the changes that have come to Washington since, as a functioning system of government turned into a scorched-earth battleground: the flood of money that means leaders now spend endless hours schmoozing donors instead of solving problems; the relentlessness of the 24-hour media spotlight; the ideological segregation that turned the once-common categories of conservative Democrat and liberal Republican into extinct species.

Our meeting comes just a few months after Donald Trump's arrival in the White House, and his appointment of top EPA officials bent on eviscerating the agency. The relish with which the administration has begun unraveling decades of environmental protections pains Jorling. Is it upsetting, I ask, to see the party of senators like Howard Baker and John Sherman Cooper led by those determined to undo so much of their work? His answer is piercing: "It eats at your soul." As a teacher, he says, "you can't discourage the next generation, you can't make it all gloom and doom. But it's getting harder" to find cause for hope.

The assault on the rules that safeguard public health has been multipronged, and devastating. Under Trump, the EPA's sworn enemies run it, men and women who spent decades doing the bidding of oil, coal, and chemical companies in charge of the agency meant to protect Americans from those industries' worst depredations.

Central to the attack has been a sidelining of the science that long guided EPA decision making. The Trump team pushed to restrict the kinds of studies the agency can consider, claiming transparency demanded the rejection of those built on subjects' confidential health information. It's a spurious argument intended to put many of the most important findings on air pollution's effects out of bounds. His EPA also rewrote the requirements for membership on its science advisory panels to exclude many scientists, claiming conflicts of interest disqualified those who'd gotten research grants from the agency (but not the executives of regulated companies, who took seats formerly held by toxicologists and epidemiologists). One new adviser believes, in the face of voluminous evidence to the contrary, that "modern air is a little too clean for optimum health."[29]

Steep cutbacks are another threat: Waves of specialists have left the agency without being replaced, hollowing out the expertise needed for meaningful enforcement. The Trump administration has taken aim at rules on power plants, on airborne toxins, on cars, and more. It has slow-walked mandated anti-pollution actions, expedited favors for industry, and created endless loopholes, while running roughshod over legal requirements by simply ignoring rules it can't get rid of. The consequences for America's air, and the health of those who breathe it, are predictable.

Despite it all, though, the Clean Air Act still stands, more powerful than the whim of one man, more enduring than one administration. Stronger even than the money-gorged industries determined to undermine it. Waiting for a president who understands that respecting and enforcing the law of the land is not optional.

* * *

Paul Billings knows that well. An air quality advocate at the American Lung Association, he's spent his life fighting for the same cause that drove Leon. When we meet in his office a few blocks from the White House, I'm struck by his resemblance to an old photo of his father— both with round, boyish faces, Paul's hair tinged with the same red Leon was once teased for. He thinks about his dad a lot, about the generosity beneath his prickliness, and the work he did "to forge a law that really is the landmark public health law of the twentieth century, saved millions of lives, changed the world and really changed the way that government approached public health problems."

For decades now, Paul Billings has been raising his voice in defense of the principle his father helped write into the statute books: that human beings are entitled to breathe air that doesn't make them sick. Of course, polluting industries would prefer to save money by pouring their filth into the air, he says. But "it's as simple as the kindergarten rule: 'You make the mess, you clean it up.'"

He didn't set out to carry on this fight. As a young man, he figured the job would have been done by now. But while he's clear-eyed about the forces that have always been arrayed against the Clean Air Act, he can also see the decades of progress it's brought: almost 75 percent less pollution emitted in the United States today than in 1970.[30] "A girl on a soccer field in suburban Maryland who doesn't have an asthma attack," he tells me. "That's a win. The fact people aren't dying at the same levels as they were, that's a win."

But it's not enough, in Paul Billings's view, that America's air is better than it used to be. It ought to be so clean it doesn't kill any of us. Over the decades, though, as the act has done its work and the dire crisis slowly eased, a new challenge has emerged. Part of the law's resilience came from its solid design. Its other crucial pillar is strong public support. But while nothing is more fundamental than the air we breathe, it's also true that nothing is easier to forget once you have it. I've experienced that myself—when London gets a clean day, yesterday's pollution drops right out of mind. So the political imperative Richard Nixon was responding to when he signed Ed Muskie's bill has lost some of its urgency, and the Clean Air Act may be at risk of succumbing to its own success.

Which means there is a danger now of going backward, reversing the progress Ed Muskie and Howard Baker and John Sherman Cooper set in motion. Paul Billings is among those fighting to protect it, but sometimes the arguments are a hard sell. "Leon and I had a picture of Beijing," Tom Jorling tells me; "our theme was, 'But for the Clean Air Act, there is the U.S.'" That road not taken is real, but what-ifs rarely capture the imagination. It's easy, Tom says, to imagine things have always been as they are now, to forget how long the struggle has been. As for his own role, and that of the old friend he's lost, he tells me he's "quietly proud." But these days, his pride is tempered by fear of what comes next. "Wrecking things," he muses, "is so much easier" than building them.

8

RELUCTANT INNOVATORS
Air and the Automakers

Smoke billows into the chamber, thick and hot, laden with poisons. It wafts into tiny channels, hundreds of them, a honeycomb of passageways. Walls big enough to cover two football fields,[1] folded together into a space hardly bigger than a toaster. Their surface dotted with tiny flecks of precious metal, extraordinary materials found in just a few spots on earth. The fragments coat the walls, layer upon layer. Platinum. Palladium. Rhodium. Metals, dug from deep underground, with the power to wrench apart the smoke's components, and reshuffle the poisons into something new. Conditions must be perfect for them to do their work, so sensors monitor oxygen levels and temperature. Over and over again, the exhaust's molecules—hydrocarbons, carbon monoxide, nitrogen oxides—land on the metals, bond with them. Atoms separate from one another, then recombine. Oxygen pulled away from some clusters, added to others. Transforming them into something new. Harmless. They separate now from the metallic catalysts and drift away, into the air.

It's an oddly flattened cylinder, hidden in the tangle of a car's undercarriage. Often caked with grime, and not particularly recognizable anyway to those of us unfamiliar with the inner workings of the automobiles we rely on to take us where we need to go. But the catalytic converter has an importance that vastly exceeds its renown. Indeed, it must be among the technological marvels that have done humanity more good with less glory than any other, saving lives every day in utter obscurity.

It was an invention with many fathers, not only engineers and chemists, but politicians and businessmen too. Among the first with a claim on paternity was Eugene Houdry. A Frenchman with a passion for race cars and a hero's record as a World War I tank commander, he revolutionized the oil industry with a new technique for processing petroleum. But in the late 1940s, growing increasingly concerned about what the fuels he'd been working with were doing to the air, he shifted his focus to treating what was coming out of automobile engines instead of what went into them.

He came at the problem from a different angle than others had. To his mind, the most promising solution wouldn't involve rejiggering engines or putting physical filters into tailpipes. He wanted instead to reorganize exhaust's molecules to transform its hazardous components into harmless ones. He'd used chemical triggers—catalysts—to achieve something similar in his breakthrough work with oil, and he thought he could do it again. So, having emigrated to America, he started a company outside Philadelphia, named it Oxy-Catalyst, and got to work.

Part of the reason running an engine put poison into the air, Houdry could see, was that the burning happening inside was never complete. So the waste products of unfinished combustion, like hydrocarbons (tiny fragments of fuel) and carbon monoxide (the gas that kills when it leaks from a faulty heater) were left to escape into the air. The trigger that would transform them was platinum, and he spread the precious metal inside his device. By the mid-1950s, it was being installed on forklifts, and the Los Angeles air pollution office wanted to test it on cars too. In a trial, it reduced hydrocarbons—a key ingredient of L.A.'s choking smog—by 80 percent.[2]

There was one problem: The lead then added to gasoline to help engines run more smoothly was gumming up the inside of Houdry's converter after just a few thousand miles, making it all but useless. Back in the 1920s, when car companies were struggling to resolve a knotty set of technical problems known as engine knocking, a team of General Motors researchers had hit on an additive that did the trick: tetraethyl lead. GM, partly owned at the time by the chemical company DuPont, joined with Standard Oil to create the Ethyl Gasoline Corporation, securing a lucrative patent on the leaded fuel that Ethyl would soon begin to sell.

A Yale physiologist named Yandell Henderson had tested tetraethyl lead as a potential nerve agent during World War I, and when GM asked his thoughts on putting it into gasoline, he replied with alarm. "Widespread lead poisoning was almost certain to result," he warned. Later, he deemed it the "single greatest question in the field of public health that has ever faced the American public."[3]

The science was clear: Lead is a powerful neurotoxin. The threat was vividly demonstrated at a New Jersey refinery whose tetraethyl lead operation was known as "the loony gas building" because of its workers' bizarre behavior—stumbling, memory loss, explosions of rage. After an accident, dozens collapsed, suffering seizures and hallucinations; more than 30 were hospitalized and 5 died.

The companies—writing a playbook polluters would draw on for decades—attacked the science, and paid for some of their own, to argue lead's dangers were exaggerated. A Standard Oil executive even called tetraethyl lead "a gift of God." The surgeon general brushed aside concerns, too, approving Ethyl's plans, and even taking the trouble to reassure foreign governments the additive was safe, so it could be marketed overseas.[4] Ethyl, GM, and the others eschewed alternative engine knocking solutions in favor of the one that would soon earn GM $43 million in patent royalties, and Ethyl nearly twice as much in profits.

Before long, more than 90 percent of gas sold in the United States contained lead. Thanks to its patent, GM, one industry expert said, "made money on almost every gallon."[5] By the 1970s, lead was being added to nearly all the gasoline in the world.[6] As a result, it permeated the air, and made its way, too, into dust, soil, drinking water, and food. Children are particularly vulnerable to the harm lead wreaks on brains; in adults, it causes high blood pressure, strokes, and kidney damage as well.

And as Eugene Houdry tinkered with his catalytic converter, it became clear that lead was also an obstacle to the adoption of his potentially revolutionary technology. Its promise could be realized only if the toxic additive was removed from fuel. But industry didn't want to give it up, and in the 1950s, with automobiles free to spew just about anything into the air, no one was ready to force them.

It was the passage of the Clean Air Act in 1970 that would turn Houdry's idea from a pipe dream into a practical necessity, the most promising answer to a problem that the law now ordered car companies to solve. The act had given them five years to reduce by 90 percent their vehicles' emissions of both hydrocarbons and carbon monoxide, and one more to do the same for a pollutant that was tougher to manage, nitrogen dioxide. Carmakers were furious. Some of the new requirements "could prevent continued production of automobiles," warned Lee Iacocca, then Ford's vice president. "They could lead to huge increases in the price of cars. They could have tremendous impact on all of American industry and could do irreparable damage to the American economy."[7]

The goals were ambitious, to be sure. They had to be. The number of cars on the nation's roads was growing so fast that only a dramatic cleanup of their exhaust would improve air quality. What's more, Senator Muskie's subcommittee had gotten some strong hints that, despite the manufacturers' protestations, such steep reductions were well within their reach. When auto executives brought their technical experts to a meeting with the subcommittee's staff, one spoke privately about his bosses to Tom Jorling. "Don't listen to them," Jorling recalls him saying. "Whatever you tell us to do we can do." And foreign carmakers were ready to comply, promising they'd do whatever they were told.

Detroit seemed uniquely unwilling to act, more interested in innovations that improved a vehicle's looks, or the driver's experience— changes that would sell cars—than those that would protect public health. But by 1970, the companies didn't have much credibility left on Capitol Hill. They'd had to settle with the Justice Department over accusations they'd conspired to slow the development of new pollution control equipment. And they'd been crying wolf for so long—any threat of new regulation elicited warnings of economic catastrophe— that Congress had mostly stopped listening.

A new constituency had sprung up, too, in the firms eager to make catalytic converters for the car manufacturers. They sent reams of documents to the pollution subcommittee "saying, 'This is easy. We can do this,'" Tom Jorling told me. The catalyst makers were afraid speaking publicly would alienate the companies they hoped to sell to, but they were beginning to make their voices heard. Their device held world-changing potential. But it would take a powerful champion inside the auto industry itself to get it out of the lab and onto the roads.

* * *

A magazine story once said Ed Cole was one of the last Detroit auto bosses who knew how to fix his own car.[8] As president of General Motors, his job was business: pleasing customers and stockholders, looking out for the bottom line. His son David spent a lifetime in the industry, too, and he tells me there are two sorts of executives, the finance types and the engineering types. "The engineering types, they can't keep away from engineering stuff." That was his dad. Ed Cole had grown up on a farm in western Michigan, and he was practical to his bones. He'd worked on tanks in World War II, and later became chief engineer at GM's Cadillac and Chevrolet divisions.

Even from the presidential suite, he always kept up with what the company's research teams were working on. In a conservative industry, another thing distinguished him, too: He wasn't afraid to take a chance. "He had a motto that people sort of associated with him, and that was 'To hell with the status quo,'" his son says. "He fought tradition." His taste for risk would bring him to an early death later on, when the private plane he was piloting went down in bad weather.

But at the end of the 1960s, it served GM—and the country—well. Even before the Clean Air Act became law, Ed Cole could see change was coming, and unlike his competitors, he was prepared to accept it. Carmakers were going to have to clean up their vehicles, and to his mind, the catalytic converter had a big advantage over other ways of doing so. Many earlier pollution controls hampered an engine's power or mileage. A catalytic converter, though, hardly affected engine performance; it just treated the exhaust that came out. GM's engineers had been tinkering with Eugene Houdry's idea, and the company had gotten its own patent for an updated version.

The problem Houdry got stuck on hadn't gone away: the catalytic converter was useless as long as gas contained lead. Unlike the French war hero, though, Ed Cole was in a position to do something about that. As president of the nation's biggest carmaker, he could single-handedly create the demand that would enable—indeed, force—oil companies to bring unleaded fuel to filling stations across America.

In January 1970, Cole spoke at a Society of Automotive Engineers banquet in Detroit, to an audience of auto and oil executives. He wanted to move forward with the catalytic converter, and he told his stunned listeners the time had come to take lead out of gasoline. "It was a very controversial announcement," his son recalls. "He said, 'We know how to fix this, and so let's do it.'" Even within GM, there was disquiet, but Ed Cole commanded the respect to push this change through.

One by one, he invited executives from all the big oil companies to come see him, as well as the two main suppliers of lead. GM had sold its stake in Ethyl, but it still wielded influence. The meetings, one GM engineer tells me, were exciting, because "forgive me—the oil companies, and especially Ethyl and DuPont, were pissed. 'Oh! What are you going to do? You're going to put us out of business. This isn't going to work.' He made it work."[9]

It was a strange position to be in, since GM had been so central to the decision to add lead to gasoline in the first place. But Ed Cole helped put the additive that was damaging children's minds on a slow march to extinction. His arm-twisting succeeded in getting the oil companies to start selling unleaded fuel, and in 1972, the EPA, using power granted by the Clean Air Act, ordered it be made widely available. Later, it required lead levels in gas to be gradually lowered, then outlawed it entirely (Ethyl fought every step). As concentrations in fuel and the air fell, scientists measured an invisible change occurring inside Americans' bodies: Average lead levels in children's blood fell from 0.16 to 0.03 parts per million. That's a big enough drop to spare a child the loss of two to four IQ points.[10]

Despite Cole's embrace of the catalytic converter, he didn't want his company forced to move too fast. So as soon as President Nixon signed the new law, GM and the other carmakers pressed for more time, deriding the new requirements as impossible to meet. Nixon had put William Ruckelshaus, a straight-arrow Indiana Republican, in charge of the EPA, and it fell to him to rule on the extension. The carmakers made their case in a packed auditorium. Two long tables faced the cameras and the crowd, one for Ruckelshaus and his staff, the other for the parade of witnesses. The battle lines were clear. Auto companies said catalysts were unworkable, and catalyst companies said they were ready to go. A Chrysler executive held up a melted hunk of metal to demonstrate the purported danger.[11] Ruckelshaus listened, then reached an answer: No. The carmakers would have to live with the deadlines the law had set.

Like nearly every decision his young agency would make, it was challenged in court. A judge ordered Ruckelshaus to reconsider, and he eventually offered a compromise: a one-year delay, with interim improvements. The companies kept fighting. Chrysler bought a full-page newspaper ad claiming, "There is no scientific evidence showing a threat to health from automotive emissions in the normal, average air you breathe."[12] Lee Iacocca, by then Ford's president, said getting rid of smog, at huge economic cost, would spare the average American just one day of coughing every 33 years.[13]

Congress would later extend the deadline further. But the die had been cast. In the end, the companies did what they had to do. They didn't meet the goals as quickly as Senators Baker and Muskie had envisioned. But by the mid-'70s, the vast majority of cars sold in America carried catalytic converters. Europe was slower, but gradually, it changed too.

* * *

One thorny problem remained. The catalytic converter worked well on two of the pollutants Congress had targeted, hydrocarbons and carbon monoxide. But eliminating the third, nitrogen dioxide, was harder, partly because it required the opposite chemical reaction. Adding oxygen gets rid of hydrocarbons and carbon monoxide, but it must be removed to reduce NOx.

John Mooney figured there had to be a way. He was a chemical engineer at a mineral refining company called Engelhard, and as the

1970s wore on, he calibrated and recalibrated the engine in the corner of his lab in Newark, New Jersey, and tested one precious metal after another in a new kind of catalytic converter. Each evening around 6:30, he'd walk into Carl Keith's office. Keith was his boss, the man who had to make the case to management that it was worth sticking with this difficult project—that the market, if they could solve the nitrogen dioxide problem, would be huge.[14]

Mooney would describe what he'd found that day, and Keith would toss out ideas for the next round of experimenting. For the new converter to work, conditions inside it had to be tightly controlled. The company sent Mooney around the world to confer with auto engineers in Japan, Britain, Sweden, and Germany. The onboard computers just being developed, they realized, could monitor and adjust oxygen levels, and it was that breakthrough that made NOx reduction possible. John Mooney and Carl Keith's new device, able to treat all the required pollutants, became known as the three-way catalytic converter, and it soon replaced the earlier devices. (Although sadly, it's not suited to diesels, which must use a separate system to treat NOx.)

The Ethyl Corporation had pushed lead overseas after American regulators banned it at home. So, late in his career, Mooney helped Asian and African nations remove it from gasoline so they could use catalytic converters too. Today, the scourge of leaded gas is gone nearly everywhere. Globally, its disappearance saves more than 1.2 million lives, and $2.4 trillion, every single year. Incredibly, researchers found it has also prevented nearly 60 million crimes—a result of the toxin's link to aggressive behavior.[15] It's just one-half of an extraordinary dual achievement: the elimination of a brain-altering poison, and the use, all around the world now, of the device mechanics call a "cat," delivering cleaner air to billions of people.

John Mooney is proud to have played a part. His old boss, Carl Keith, has died, but I reach Mooney on the phone at his home in New Jersey, not far from where my parents live. His voice is thick with age, and his memory of some of the long-ago details has dimmed. But the pride in his work's far-reaching impact burns bright. "Yep. My chest is sticking out right at the moment," the old engineer tells me. "I have a nice smile on my face."

* * *

Joe Colucci had a close-up view of the catalytic converter's long, slow birth too. He grew up in the Brooklyn workshop where his dad made seat covers for cars, then found his way back to the industry as an adult, rising to run one of GM's big research departments. When I call him at his home in Ohio, he tells me Ed Cole was one of his heroes. In the 1980s, long after Cole's lead decision, Colucci would be part of another big change, pushing the oil companies to make gasoline cleaner still, reducing sulfur and other pollutants.

He knew such progress didn't come on its own. At GM, "it was taken for granted that we have met the enemy and he is the EPA," Colucci says. "It was instilled in us, it was just the ethic. Because they're going to force us to spend more money." When the laws and rules came through, the company mostly did what it had to, but the remarkable improvements in America's air wouldn't have happened had the industry been left to its own devices. "I can't think of where this country would be now if we didn't have the state of California and the EPA beating us over the head to get these things done," he tells me.

Indeed, California has long played a crucial role in America's pollution struggles, dragging the nation along behind it. Because it had already begun regulating car emissions when the Clean Air Act passed, the law allowed it to continue doing so. No other state can set its own rules, but they can sign on to California's, and the size of that collective market has given Sacramento great power.

So while it's true that Ed Cole had the foresight to wrench his industry forward, it's also fair to say he did it only because someone forced him to. That's been the carmakers' story for decades, and it's a story still unfolding, as they cling to the past in the face of a new wave of change. Yes, it's a story of innovation, of technological advance. But innovation that came so much more slowly than it might have— human beings and the planet choking in the meantime—and was so much less bold than it could have been. Change that, when you come down to it, had to be dragged out of an industry that directed vast resources at fighting the future instead of embracing it. Joe Colucci still remembers his GM years fondly, so he puts it differently. "Yeah, we had some warts. Everybody has warts, but we did a hell of a lot to make society better. Maybe it was required," he laughs. "But we still did it."

* * *

Sometimes, carmakers don't just resist the rules, they deliberately break them, as John German discovered. German is in Washington visiting a new baby in the family, his wife's grandson. He looks the part of the no-nonsense midwesterner, with wavy white hair, a beige button-down shirt, and dark green trousers covered in loops and pockets, and he seems a little out of place in the hip Dupont Circle café where we meet. So do I, truth be told, although I'll blame the suitcase I'm lugging, having just arrived in town, rather than any innate lack of hipness. German grew up in Detroit—"everybody was in the auto industry"—and has spent his life in Michigan. He speaks quickly, pausing for sips of hot chocolate, as he describes the discovery that unexpectedly catapulted him and the modest research group he works for into the headlines.

German hadn't been looking to make a splash when he and his colleagues at the International Council on Clean Transportation commissioned an examination of pollution from diesel cars, comparing what came out of their tailpipes during the lab tests required by law with emissions on the road, under real driving conditions. They just wanted to tie up the last loose ends in a big report, and thought the research would give them something positive to say about diesel, tips for Europe from America's experience in getting the dirty fuel to run a little cleaner.

Needless to say, that wasn't how it turned out. They chose a Volkswagen Jetta as their first test subject, and VW's Passat next. California regulators agreed to do the routine certification test for them, and the council hired West Virginia University researchers to drive the same cars through cities, along highways, and into the mountains, using equipment that tests exhaust straight from the tailpipe.

It was clear right away that something was off. At first, German wondered if the cars might be malfunctioning, and he asked if a dashboard light had come on. That didn't really make sense, though—they had just passed the regulators' test. His partners thought there might be a problem with their equipment, and they recalibrated it again and again. But the results didn't change. Nitrogen oxide pollution from the Jetta's tailpipe was 15 times the allowed limit, shooting up to 35 times under some conditions; the Passat varied between 5 and 20 times the limit.

John German had been around the auto industry all his life, so he had a pretty good idea what was going on. This had to be a "defeat device"—a deliberate effort to evade the rules. "It was just so outrageous. If they were like 3 to 5 times the standards, you could say, 'Oh, maybe they're having much higher NOx emissions because of the high loads,'" or some other external factor. "But when it's 15 to 30 times the standards, there is no other explanation," he says. "It's a malfunction or it's a defeat device. There's nothing else that could possibly get anywhere close."

He wasn't ready to level such a serious accusation against a huge company like Volkswagen, so John German kept quiet while the research moved forward. Much later, his boss was surprised to learn how early he had suspected the truth. "He said, 'You knew there was a defeat device? Why didn't you tell me?'" The answer was simple. "We're an $8 million organization. VW could have squashed us like a bug."

German and his colleagues pressed ahead with their work, and when the study was finished, they posted it online. That was May 2014. He was still nervous, so the council didn't issue a press release, nor did the report name the manufacturer. As a courtesy, he sent a copy to someone he knew at Volkswagen, noting, "By the way, Vehicles A and B are yours." German's group also forwarded the findings to the EPA and California's Air Resources Board. "We were definitely scared. We wanted EPA and CARB to take over." Afterward, he'd email the agencies now and then. No one replied, and having spent more than 13 years at the EPA himself, he knew what that meant.

The regulators were investigating. And while they struggled to determine what was causing the discrepancy between pollution in the lab and on the road, VW executives quietly debated their next move. "It should first be decided whether we are honest," one wrote to a colleague, prosecutors later alleged. The answer the company settled on would soon become abundantly clear. After months of foot-dragging, VW promised to remedy the problem, which it blamed on a technical glitch. It began recalling cars, updating the software in hundreds of thousands of them.[16]

Months later, California ran new tests. Emissions were still far over the limit. Now regulators wanted to see the software controlling the vehicles' pollution systems. And they leveled an extraordinary threat to get it: If Volkswagen didn't turn over the code, it wouldn't get the approvals it needed to sell cars in California and a dozen states that used its standards. The EPA threatened to withhold certification for the entire U.S. market.[17] "That," John German says, "was when VW came clean."

Dieselgate, as it became known, exploded into one of the biggest corporate scandals in history. Over nearly a decade, Volkswagen acknowledged, it had embedded defeat devices in 11 million cars, mostly in Europe, about 600,000 in the United States. The software detected when emissions tests were being run, and pollution controls worked fine under those circumstances. But outside the lab, the controls were switched off or turned way down, and NOx levels shot up as high as 40 times the legal limit. With mind-boggling gall, Volkswagen had even used the software update it was forced to carry out to improve cars' ability to detect when they were being tested.[18] It wasn't the first to get caught using a defeat device, but the boldness of VW's cheating and subsequent concealment were stunning even in an industry with such a rich history of hostility to regulation.

When the news broke, German was overwhelmed by reporters' calls: "It was absolutely insane," he recalls. As the day finally ended, he breathed a sigh of relief. "It was like, 'Oh, thank God this is over.' The next day was twice as bad. I did almost nothing else for two or three months, 12 hours a day."

At first, the company blamed "the grave errors of very few," but it quickly became clear the evasion was too systematic to have been carried out by a handful of rogue engineers. In the United States, VW eventually agreed to pay close to $30 billion in penalties and settlements, compensating drivers and recalling and buying back cars. CEO Martin Winterkorn said he hadn't known of the cheating, but he resigned and was later charged criminally (although he's unlikely to be extradited). Other high-ranking executives were indicted, too—one was apprehended on vacation in Florida—and guilty pleas followed. The company also pleaded guilty, to charges including conspiracy to violate the Clean Air Act and obstruction of justice. "Volkswagen obfuscated, they denied and they ultimately lied," said then–Attorney General Loretta Lynch.

That's what really gets John German, and what made VW's legal problems so much worse. Even after getting caught, instead of removing the defeat devices, "they spent 16 months lying to the agencies." That, he says, "is the part I find so incomprehensible." What's clearer is why they cheated in the first place. It saved money. Diesels are harder to clean up than gas cars, and the systems that do so are more expensive. They sometimes hamper power and fuel efficiency. The defeat devices meant VW could use a cheaper approach instead. "Because it didn't have to work."

And, as it turned out, VW wasn't the only one evading the law. Less flagrantly, but to similar effect, the vast majority of diesel cars were making a mockery of emissions rules. In the wake of the revelations in America, European governments road-tested not just Volkswagens, but other big brands too. Germany found all but 3 of 53 models exceeded NOx limits, the worst by a factor of 18.[19] In London, the testing firm Emissions Analytics found 97 percent of more than 250 diesel models were in violation; a quarter produced NOx at six times the limit.[20] "As the data kept coming in, our jaws just kept dropping. Because it is just so systematic, and so widespread," John German says. "VW isn't even in the worst half of the manufacturers." With a few honorable exceptions, "everybody's doing it."

* * *

In the United States, where fewer than 2 percent of cars are diesel,[21] the rule breaking had an impact, to be sure. But the health consequences have been far more severe in Europe, where drivers had been encouraged for years to buy diesel cars and they accounted, when the scandal broke, for more than half of all sales. In 2015 alone, failure to comply with the rules caused 6,800 early deaths, one study found.[22] To put it more plainly, tens of thousands of people had died because carmakers felt so free, for so long, to flout the law.

Of course, the painful light cast by the scandal didn't just expose corporate wrongdoing. It also made visible a failure that, to me at least, is just as distressing. Across Europe, governments responsible for enforcing the law and protecting their people's health had utterly neglected to do so. Just as much as VW and the other cheating companies, the revelations implicate the officials who didn't catch them. The fact of the matter, John German explains to me, is that European air quality regulators don't have the muscle or the resources their American counterparts have long possessed. European countries have never built the enforcement capability needed to give their pollution rules teeth. Governments, he says, "don't seem to be able to do anything about it, in most cases don't even seem to want to do anything."

The European arm of John German's group had reported back in 2014 that cars they'd road-tested averaged seven times the EU's NOx limit. But Europe took action only after U.S. regulators' response made the cheating impossible to ignore. Even then, it moved far more slowly, and less aggressively, than American officials had. As a consequence of that lax approach, and the legal gray areas it's created, VW, even after confessing wrongdoing in America, maintained that its emissions-manipulating software was legal in Europe; it admitted its actions, but denied they were against the law. With 8.5 million VWs carrying the rule-evading code in Europe—14 times as many as in the United States—the financial consequences of any finding to the contrary could be ruinous.

Americans are accustomed to thinking of themselves as environmental laggards compared with Europeans. Rightly so, much of the time. Our average carbon footprint is double a European's,[23] and we shout if gas hits a fraction of what it costs in France or Germany. But the Clean Air Act and the strength of the agency enforcing it have given Americans far healthier air than their allies across the Atlantic.

The EPA has, over the years, built up tremendous legal and technical expertise. At least until the evisceration of the Trump years, it was known for its diligence in supplementing regulations with circulars and advisories that precisely defined every term, clarifying ambiguities and laying out what was allowed and what was not. The result was a system that, if not watertight, was a lot less leaky than elsewhere. In Europe, while the rules may look similar, no one goes to the trouble of making clear exactly what they mean, so polluters provide their own interpretations. Its atrocious air offers a cautionary tale those undermining American regulation would do well to heed.

Years on now from his brush with fame, there's one thing that still bothers John German. He sometimes hears skeptics suggest VW's wrongdoing didn't do any harm. "That," he says, "really sets me off." Unlike scandals involving brakes or airbags, a driver wasn't personally endangered by this malfeasance, and of course it's impossible to say exactly who was. But, with everything we know about pollution's dangers, it's clear spewing toxic exhaust into air human beings must breathe brings deadly consequences. Selling cars that evade pollution rules, German says emphatically, "is not a victimless crime."

* * *

As I wheel my suitcase out of that Washington café, past a restaurant where I waitressed one summer in college, I realize John German has given me, finally, a full understanding of the pollution that has obsessed and infuriated and terrified me all these years in London. I see now that blame doesn't lie just with the misguided turn-of-the-century decision to nudge drivers toward diesel. Nor only with car companies' rule breaking. The failure of so many governments to enforce the law is the missing third piece of this puzzle. What I understand now is that the people we entrusted with the power to protect us essentially decided not to bother. Instead, they have allowed automakers to spew whatever they want into our air.

That, at bottom, is the explanation for the filth that leaves grit on my teeth and a sour taste in my mouth, and puts me and my family and my neighbors at risk of heart troubles, dementia, early death in many forms. Together, the mistake and the cheating and the negligence are why the streets we walk every day in London are fouled by noxious fumes.

How could this have happened? In countries that are among the wealthiest in the world, on a continent whose name is a byword, elsewhere, for environmental progressivism? I still can't quite fathom it, and I want to understand how this gap between rules on paper and reality in the world could have grown so wide. So, not long after meeting John German, I'm in a conference room in Berlin, as music from a guitarist in a busy square below wafts through open windows. This is the European office of the International Council on Clean Transportation, whose American arm German works for, and I'm here to see his across-the-ocean colleague Peter Mock, a younger man than I expect, with a silvery shirt and slightly spiky dark hair.

As Mock speaks, I begin to absorb the particulars of Europe's stunning failure. Like much in the world of air pollution, the details sound dull and bureaucratic, but they carry, quite literally, life-and-death consequences. It starts with an enforcement structure that almost seems designed to let violators through. The European Commission sets the rules on how much pollution a car is allowed to produce. But the job of enforcing them falls not to Brussels, but to national governments. And a car company preparing to release a new model can choose which country certifies it; every EU nation must then honor the approval. A savvy automaker opts for someplace it provides lots of jobs, where officials are likely to be pliant.

The national enforcement agencies, for their part, are generally understaffed, poorly funded, and lacking technical expertise. In Germany, I learn, responsibility has long been divided between two agencies. While the Environment Agency gathers emissions data, Transport certifies new cars. But Transport is also the department that tends governmental relationships with the country's powerful auto industry, and the desire to keep those ties warm doesn't mesh comfortably with tough enforcement.

Britain is an exception, but in most nations these weak agencies don't even test cars themselves. About a dozen individual vehicles must be checked before a new model is approved, and the tests are often run by outside contractors. Those companies rely on the carmakers' business, so they have good reason not to provide unpleasant results. When they're done, the manufacturers hand the paperwork to regulators, and Mock says results are usually accepted with little question.

What's more, the specifics of the tests—speed, acceleration, and such—are publicly available. So a manufacturer can build its cars to produce little pollution under those particular conditions and lots the rest of the time. This is the key to a question that's been nagging at me. I know not just Volkswagens, but many diesel brands, shatter emissions limits. Yet most companies don't have VW's legal troubles. How are they getting away with making equally dirty cars, when everyone knows they're breaking the rules? Like exam-prepping pupils who focus, laser-like, on the facts that will win good grades, it turns out many manufacturers design pollution systems just to pass the test.

There's another route those companies take, too: programming pollution controls to turn off when the weather's too hot or too cold, when a car's just being started or is speeding up or slowing down or climbing a hill, conditions they frame as extraordinary but that account for a big chunk of driving time. If challenged, the companies can cite a legal loophole, claiming the switching off is necessary to protect engines. Such tricks occupy murky territory in which manufacturers don't declare them and authorities don't rule on their permissibility.

Now, at last, European regulators have begun requiring cars to be tested on the road, not just in the lab. Many tout the new approach as the one that will finally bring automakers to heel, and force them to sell cleaner cars. If the problem is the discrepancy between lab tests and on-the-road performance, the thinking goes, then moving tests out into the world is the solution.

But the real problem, to my eyes, is even bigger. Because it seems clear the flaws in European nations' enforcement are more fundamental than the particulars of one testing method. The problem is the system itself, a system riddled with weakness and ripe for abuse. Politicians have begun, post-Dieselgate, to tighten it, but it remains a system designed under the gaze—and the lobbying pressure—of a powerful industry.

Germany's automakers wield a special influence, a potent symbol of that country's manufacturing might. Their web of relationships reaches government's highest levels, and Germany's power in the EU means that influence is felt across the continent, as Berlin pushes for weaker pollution rules and lax enforcement. Sometimes it even stretches over an ocean: Angela Merkel once complained to California regulators that their tough pollution rules were hurting her country's manufacturers. "She was there, it seemed, as spokeswoman for the auto industry," the state's top air official would recall.[24] Asked during testimony in the Bundestag why Americans, not Germans, had uncovered the VW scandal, the best Europe's most powerful woman could offer was "I don't have any explanation."[25]

The industry, it's clear, has often succeeded in persuading those in power to put concerns about its profits above the public's health. The contrast with the priorities Ed Muskie set in the United States is stark, and so, as a result, has been the outcome. On-the-road testing may be an improvement, but the hopes that it will be the panacea that finally delivers clean air seem overblown. With an enforcement system this ineffective, the governments of Europe have essentially let car companies police themselves. The consequences, awful as they've been, are unsurprising.

In fact, I learn to my astonishment that some in power knew about the consequences all along. I speak by phone to Martin Schmied, an official at Germany's Federal Environment Agency. His department, he tells me, has been taking cars on the road for 25 years to measure emissions—and publishing the results. Year after year, they found diesels producing NOx at many times the legal maximum; six times, in one recent test. Flabbergasted, I ask him to clarify: Germany's government, and anyone who read its public reports, has known for decades that automakers were flouting the rules? Schmied responds with equanimity. As long as emissions went down when limits were tightened, his department didn't mind they were many times higher than allowed. "We publish this data," he says. "In principle, this is nothing new."

So Germany knew. Perhaps other governments did too. Many of their people, though, did not. I certainly didn't. Nor did the buyers of millions of diesel cars. Nor the hundreds of millions of people who breathe the air they taint, trusting for so long that companies were following the law—and that governments would catch them if they didn't. I think of Sir David King, Tony Blair's scientific adviser, who told me he gave his support to the tax changes that would put so many diesels onto British roads because he believed they would meet emissions limits. And I wonder whether he, and the others who shared that view, might have seen things differently—might not have set Britain on that deadly course—had they known government statistics being churned out in Germany were telling such a dramatically different tale.

The diesel cheating scandal, it strikes me, is in some sense a failure of innovation, yet another symptom of car companies' perennial desire to stick with what they know, with the cars that reliably deliver profits. That caution is surely at the root of why European manufacturers pushed governments looking to shrink carbon footprints to turn to diesel rather than, for example, hybrids like those Honda and Toyota had already put on roads by the late 1990s. With their vast resources and the marketing muscle to bring consumers along, who knows what VW and the others could have come up with. We have all paid the price for their decision not to try.

* * *

Today, glimmers of a different future are in sight, as electric cars begin moving from niche to mainstream. There are challenges, to be sure: the need for better batteries, more charging points, and enough power to keep cars supplied. But those are obstacles that can be overcome, and the technology is advancing quickly.

The stakes are higher than ever. Today's cars are, by some measures, 99 percent cleaner than the barely regulated ones of 1970.[26] But while governments have taken aim at the pollutants that harm our bodies, they have hardly begun to target the one that's warming our planet. Until recently, no auto regulation sought to reduce carbon dioxide. So it has climbed along with the number of cars on the road, the miles they drive, and the gallons of fuel they burn.

Today, electric vehicles look like the best way to slash both sorts of pollution, another place where the goals of a healthy climate and healthy bodies converge. Electricity by itself is no guarantee of climate friendliness. But it's a necessary prerequisite to powering cars from clean sources like wind and solar.

Electric cars are not a cure-all. While they don't create exhaust, their brakes and tires give off tiny, toxic particles as they wear. The energy needed to manufacture them, the raw materials used in their bodies and batteries, will be an unsustainable burden on our groaning Earth if car ownership keeps increasing.

For now, that relentless rise frames everything. The number of vehicles in America has more than tripled since 1960;[27] in England, there's one car for every two people.[28] And the biggest growth is now in developing nations like India and China. If they follow the path we've taken, the world could go from about a billion cars today[29] to more than 3 billion by 2050.[30] What's really needed is not just a slowing of that growth, but fewer cars altogether, of any sort. It's a goal that's reachable if we reorganize the places we live to be denser, more pedestrian and bike friendly, with public transportation—and newer options like car sharing—that are convenient and affordable.

Still, cleaning up the vehicles we do drive is crucial. As in the past, the best hope comes from companies willing to put in the money and brainpower needed to do it. And as ever, a lot of powerful players are deeply invested in the old ways of doing things, so progress has sometimes been grudging.

But this time, there are some important new forces at work. One of the biggest is China. We'll explore the reasons behind its push toward a cleaner future later. For now, it's enough to know the country's leaders have recognized the urgency of confronting their pollution problem, and they are eager to dominate the industries needed to do so, electric cars very much included.

China's hunger for clean cars—along with its willingness to put big money into its top priorities—is transforming the sector in ways likely to affect us all. It is by far the world's largest market for cars, and its demand for electric ones is ramping up fast, so it's now the biggest buyer of those too. The government is aiming to get millions more onto roads, offering rebates to drivers who buy electric and telling big multinational automakers that if they want to do business in China, they have to hit ambitious targets for low- and zero-emission models. That aggressive push is sure to accelerate trends already under way globally: falling prices and advancing technology, particularly better batteries that increase range and charging speed.

It's not just cars China wants to clean up. Buses may be less glamorous, but because they're urban workhorses—on the road in densely populated areas for many hours each day—slashing their emissions brings big health benefits. In Shenzhen, buses in the early 2010s were half a percent of traffic but created 20 percent of pollution. With astonishing rapidity, the city became the first to electrify its entire fleet, more than 16,000 buses. The rest of the country is following suit, with hundreds of thousands more electric buses coming into service. As with cars, China's huge demand is creating economies of scale likely to push prices down elsewhere. "The market needs a kick in the pants to really get moving," one writer opined. "And it looks like China is providing it."[31]

Of course, it's not the only one demanding cleaner vehicles. In the United States, California is out front, pushing to get millions of zero-emissions cars onto roads. Norway is another leader, and Europe more broadly is also taking to electric cars—because of drivers' anger over the diesel scandal, a wider understanding of air pollution's threats, and the need to confront climate change.

Carmakers seem to be responding—some faster than others—by rolling out new electric and hybrid models and promising more. Funnily enough, one of those jumping into this new market is VW, desperate to erase the stain of its cheating and experienced at making affordable mass-market cars.

Electric vehicles are still just a tiny fraction of the total, but their share is growing. There are other options, too—hydrogen is one, perhaps a little further in the future. So many other industries have been transformed in recent years—TV and music, news and retail—it would hardly be surprising to see this one forced to change too. It's too soon to say whether the big, long-standing players will move fast enough, but it's becoming clear that if they don't want to do the job, there are others who will.

*   *   *

It feels fitting that, to get a glimpse of a cleaner future, I must barrel down a highway crowded, at the tail end of a Northern California rush hour, with the gas-guzzling SUVs and big diesel trucks of the present. I'm heading to Tesla's factory in Fremont, on the edge of Silicon Valley. Indeed, it feels inside as much like a tech operation as a traditional manufacturing one, not postindustrial, but post–fossil fuel, post-dirty. The vast building, a mile and a quarter long, is surrounded by acres of parking. I've been warned to allow at least 15 minutes to find a spot, a wise suggestion, and except for a few plug-ins at the front, nearly all the cars here are the traditional kind.

Inside, a klatch of Tesla enthusiasts has gathered for this morning's tour. As our guide, Kim, in black slacks and T-shirt, her gray hair long and loose, leads us toward the factory floor, she begins a well-practiced patter, mixing facts and figures with jokes and shout-outs to fellow Tesla owners. Her enthusiasm seems genuine. "Each and every day you drive your Tesla, each and every mile, you are helping to save the world," she tells us.

In a small demonstration area, Kim passes around some of the materials used to make these vehicles. First comes a small cylindrical battery, the size of two or three AAs; about 7,000 of them, she explains, are packed together into the pale green, flat-bottomed case that forms the undercarriage of a nearby display car. Next comes a metal ingot, then a tub of black plastic beads that remind me of the more colorful ones my artsy daughter arranges into heart and star shapes for me to melt together with an iron. When we climb into a long trolley, Kim slips on a headset and gets behind the wheel.

The factory is white and airy, flooded with natural light. Even more striking than its cleanliness is the quiet. There's a low hum of machinery, but aside from the occasional beep or clank, you could easily take a nap or read a book here. There are no roaring engines or unpleasant fumes. No soot, no grime. Aside from their heavy work shoes, most of the employees I see, in jeans or black pants with T-shirts and baseball hats, could be behind the counter at an electronics store, or in the office of a tech start-up. They whisk past us on bicycles or scooters, and walk in lanes painted onto the shiny floor. One has hopped off his bike and is scrolling through his phone.

As Kim drives and talks, I gaze at a sea of metal parts, stacked on shelves in rows hundreds of feet long. Some are recognizable as fenders or doors; others have shapes whose meaning I can't discern. A press as big as a small building turns giant rolls of aluminum into body panels; Kim says its foundation extends three stories downward too. Several workers inspect the pieces emerging on a conveyor belt. At many stations, no one is present, just red robotic arms, quietly sliding back and forth, up and down, spinning and turning and lifting, riveting and welding. But while the employees are scattered, there are a quite a lot of them, all told. Many sit or stand at computers in office-like clusters of desks and tables that open onto the larger work floor. We pass an upscale cafeteria with a salad bar and coffee counter; it's also open to the factory floor, and it's filled with workers chatting and eating.

As we drive on, I begin to see the shells of cars, scores of them lined up in rows. In the "hang-on area," doors are being installed, and seats are stacked nearby. Tires, too, then more car bodies, freshly painted ones. As Kim bids us farewell—"Hopefully I showed a lot of the magic and the mystery around your car," she says—I finally realize what's missing, the part I haven't seen here, one so familiar it's taken me until now to clock its absence. There are no exhaust pipes on these cars.

Tesla—with its sleek style and big ambitions, its well-publicized troubles, and a CEO, Elon Musk, whom one columnist dubbed "the id of tech"[32]—has taken on outsized symbolism as the representative of an industry hoping to jump from its infancy straight into adolescence and beyond. Its cars drive smoothly, require little maintenance, and are replete with clever touches like door handles that pop out when a driver approaches and large touch screens in place of old-fashioned dashboard controls. Teslas' desirability, and the hype around them, spring from buyers' belief that they offer not just a replacement for traditional cars, but something far superior. The company has struggled to manufacture fast enough to fulfill promises to customers, and financial problems at times have made its future look cloudy. But it's clearly succeeded in providing a proof of concept, settling once and for all the question of whether electric cars can be both reliable and elegant, able to give drivers what they need and what they want. In doing so, it's prodded others to follow. And in the big picture, that matters far more than the fortunes of this one company.

Musk is a brash South African immigrant who, in many ways, epitomizes what America has always imagined itself to be: daring and hard-driving, ready to dream big and take risks. Despite his bravado and some headline-making lapses of judgment, his approach to business and engineering is a stark contrast to the can't-do foot-dragging that has always been Detroit's attitude toward change. He is sometimes compared with Apple's Steve Jobs, but, with typical tech-world self-belief, Musk sees his mission as loftier than making beautiful gadgets—he wants to put technology to work solving humanity's most pressing problems. One observer, recalling an investor's description of Silicon Valley as "a lot of big minds chasing small ideas," said that for all his flaws, Musk's "mind and ideas are big ones."[33] A provocateur who is often too eager to pick fights, he is an imperfect messenger for the new industries he champions. And some of his ideas sound outlandish—a colony on Mars; high-speed travel through underground "hyperloops." But he's made more progress than many expected toward achieving some of them. SolarCity is his bid to expand clean power use, Powerwall his push for the batteries that make renewables reliable. With Tesla, he wants to remake transportation, and he bet—correctly—that he could beat a vast but moribund industry to it.

Walking out of the factory, I see an oversized American flag billowing in the distance. Right beside it, another banner bears Chevron's familiar red-and-blue logo. You don't have to look far from Tesla's bubble to see the fossil fuel economy is still going strong. And in case

I needed another reminder that electric cars are still a tiny speck in a huge gas- and diesel-powered sea, a truck carrying a half-dozen shiny Teslas pulls up behind me as I head toward the highway. Despite its cargo, the truck is the traditional sort, dirty and lumbering. Almost certainly diesel.

So there's a long way to go before technology fulfills its promise on a scale big enough to matter. But Tesla, and others taking up the gauntlet it has thrown down, offer a peek at what's possible. Less important than whether that future is delivered by Silicon Valley or Detroit, Beijing or Wolfsburg, Germany, is that it dawns quickly. Just as the catalytic converter brought a leap long ago by rearranging hazardous fumes into something completely new, innovation today offers the hope of another transformation, a revolution that finally takes us where we need to go.

9

INCH BY INCH

L.A.'s Long Road

A gulp of air. Oxygen, nitrogen, water vapor. Life-giving. The breath enters the body, moves toward the lungs. Suddenly, though: a trigger, a change. Something is different. Wrong. Tiny contaminants, invisible in the mix of gases, graze the walls of the airways. An alien molecule reaches a receptor within the wall, then another, and another. Thousands of them. Millions. Each attaches to a receptor, binds tightly. Each receptor sends out a signal. The reactions begin, then cascade. In the walls of the airways, smooth muscle contracts. Tissue swells. The tubes narrow. A frightened gasp. Mucus pours into the passageways, heavy and thick. Usually, it protects, but now it is a threat, and the wheezing worsens, tightens, the pitch rising. Fear, then panic. Cold sweat. The heart races. Coughing, a spasm that won't stop. The breath is faster, but shallow, labored. A hand reaches into a pocket, a backpack. Brings something to the mouth. A pump. An inhalation. Hold it. Exhale. Then another. An easing now, as the cascade reverses. The swelling subsides. The mucus clears. Smooth muscle relaxes. The passageway opens. Air flows again. A breath, then another. The crisis past.

It started suddenly one July morning in 1943, when a stinging yellow-brown cloud engulfed downtown Los Angeles, causing panic and confusion. Eyes burned, and the air smelled like bleach. Some feared it was a Japanese chemical attack.

The city had experienced haze before, but nothing like this. And soon, foul clouds like the one that rolled in that morning were a regular occurrence, smothering Angelenos and eclipsing the sunshine they had regarded, until then, as a birthright. On bad days, schools closed, airplane pilots couldn't make out runways until seconds before landing, and the poor visibility caused car and motorcycle crashes. In fields outside town, crops withered and, slowly, groves of pepper trees and Ponderosa pines died off too. Protesters began donning gas masks to demand authorities do something. "Daylight dimout," the *Los Angeles Times* proclaimed after one episode. Within a few years, *Time* magazine had labeled the city an "airborne dump."[1]

L.A.'s smogs were unlike pollution seen elsewhere. And while it's obvious in hindsight, no one understood back then what was causing them—the heavy brown fog appeared to be coming from nowhere. At first, officials had blamed a suspected leak at a chemical plant, but shutting it down didn't help. And in 1948, Angelenos' anxiety grew when they heard about the clouds of pollution that had killed 20 people and sickened thousands in a Pennsylvania steel town called Donora. Donora was national news—experts later blamed emissions from a zinc plant, held in place by an unusual weather pattern—and while the deaths there sparked Americans' first widespread awareness of air pollution, Los Angeles was no closer to identifying the source of its own problem.

That finally changed early in the 1950s. Arie Haagen-Smit, a Dutch biochemist at Caltech, had put aside his work analyzing flavor compounds in pineapples to start running tests on the smog. He'd found it was mostly composed of ozone. That explained the unpleasant effects, since ground-level ozone, unlike the high-in-the-sky layer that protects us from the sun's radiation, causes damage and inflammation to the airways, triggering responses from wheezing to heart attacks. But Haagen-Smit still hadn't solved the biggest mystery. While ozone was present in large quantities in the smog, it wasn't widely emitted in L.A. So where, exactly, was all that ozone coming from?

The culprits, Haagen-Smit eventually explained, were the region's ubiquitous automobiles and the refineries producing gasoline for them. But they just released smog's raw ingredients—unburned fuel and nitrogen oxides, or NOx. The third element was the only thing Angelenos loved more than their cars: sunshine. Haagen-Smit suspected the sun was triggering a chemical reaction, prompting those airborne ingredients to interact and form ozone. And when he exposed car exhaust to light in his lab and watched the brownish mess form, he knew he was right. While the word "smog" is now used colloquially for any kind of air pollution, this is its more precise meaning: "photochemical smog," the technical name for those ugly clouds.

It was an impressive bit of detective work, but Haagen-Smit's answer was not what anyone wanted to hear. Cars had already become central to Southern Californians' widely envied way of life. And the oil and auto industries, of course, tried to discredit his findings. So he invited the public to watch him repeat his experiment, and he produced smog right in front of audiences' eyes—and noses. It was obvious now that making cars cleaner would be essential if breathable air was to be part of the cherished Los Angeles lifestyle.

* * *

California was the first to begin demanding such change, but the industry's foot-dragging meant it would take time. Decades. And while Angelenos grew accustomed to their smog, it could still come as a shock to new arrivals. Mary Nichols, for one, was disgusted the first time she laid eyes on it, in 1969, at the end of a cross-country road trip during her summer break from Yale Law School. "The air was this particularly odd orange color that I'd never seen before, and it smelled terrible," she tells me when we meet, on a scorching summer day half a lifetime later, in a sunny conference room at the state environmental building in Sacramento. Kitschy knickknacks—a plastic Buddha with a nodding head, a pair of retro-style lunchboxes, a waving, white-gloved Queen of England—dot the shelves. Nichols, in a crisp linen jacket, her silver hair cropped short, recalls thinking, "This place is really scary, it's not pleasant." The health consequences of the muck in the air weren't yet well understood, but one scientist put it to me this way: L.A.'s pollution back then was just below levels where exercising rats would die after two hours of breathing it.[2]

Despite her jarring first encounter with the city, Mary Nichols was back for good two years later, in 1971, with her husband and her new law degree. She didn't have a job yet, so after dropping him at his office each morning, she'd sometimes go exploring. The couple had moved into an Art Deco building in the hillside enclave of Los Feliz, and L.A. was beginning to grow on her, thanks in part to its contrast with snowy upstate New York, where she'd been raised. "I loved the weather, the beach, the palm trees," she wrote much later. But she still couldn't abide the air. She got depressed, then angry, about the yellow-gray mornings, the smog's metallic taste, and the way it stung her eyes.[3]

She hadn't thought much about pollution before moving to California, but Nichols grew up steeped in social consciousness. Her father, an electrical engineering professor at Cornell, was a lifelong activist, battling for a roster of causes including civil rights, labor unions, and, ahead of his time, gay rights. As mayor of Ithaca, he was one of just a handful of democratic socialists to hold office in the United States. One summer, a sharecropper put Mary up while she'd worked registering voters in Tennessee.[4]

So it felt natural to join a group of attorneys trying, from their offices on Santa Monica Boulevard, to use their legal skills to do some good. Not long after she started, the city of Riverside phoned the firm to ask for help. It lies 60 miles east of L.A., in the area known as the Inland Empire, right where the towering San Bernardino Mountains block the smog as it blows in from the coast. The filth settles there and sometimes sits for days. Often, it would upend the rhythms of ordinary life, forcing the cancellation of baseball games, track practices, even school recess.

Nichols's more senior colleagues were busy fighting nuclear power plants and new freeways, so Riverside's case landed on her desk. She quickly realized the city had little chance of winning if it sued Los Angeles, as its leaders had hoped to do, but she had another idea. She wanted to put the brand-new Clean Air Act to the test, so she persuaded Riverside to use it to demand California and the EPA come up with a plan for tackling smog. They won the case, and the judge ordered authorities to get serious about the region's pollution.

There were many more battles ahead, but it was an important moment for both the young lawyer and the young law—her first suit, and the first ever filed under the Clean Air Act. Their stories would be intertwined in the years to come, as Mary Nichols grew into one of the most important figures in the nation's long march toward healthier air. She was only in her mid-thirties when she became head of California's Air Resources Board in 1979. But the unique role the act had given California meant the agency's power—and Nichols's—would reverberate across the country.

* * *

Los Angeles is as good a place as any to see the Clean Air Act at work— the angry debates, the unsatisfying compromises, and the years of slow, grinding progress. The benefits the law delivered here, and its shortcomings, are part of a story that was unfolding across the nation, in Houston and Pittsburgh, Atlanta and Cleveland, New York and Philadelphia.

Indeed, when it comes to air pollution, L.A.'s story is America's, but more so—both the problem and the response have been sharper here than elsewhere in the country. Nature can take part of the blame for the city's particular challenges. In addition to the abundant sunshine that helps create its ozone, the mountains that surround the area, giving it its rugged beauty, also prevent pollution from floating away. Spanish explorers who, in 1542, got a glimpse of the San Pedro Bay—today the epicenter of the city's air woes—called it the Bay of the Smoke. They were probably seeing wisps from the native Tongva people's cooking fires, held in place by a low, invisible ceiling called the inversion layer, created when cool air coming off the ocean is pressed downward by a blanket of warmth.

Of course, there was a lot more man-made muck around by the 1970s. So Mary Nichols's agency, guided by science and working with the EPA and local regulators, began a systematic, step-by-step clampdown on polluters. The tool she used was unglamorous, but effective: regulation, the sometimes byzantine, always arduous process of rule making and enforcing. There's a reason, of course, that regulation has long been one of America's most bitter political fault lines. It is, at its essence, about power. The power, in this case, to decide what was in the air Angelenos would breathe.

Every step was hard fought. Once, tanker trucks surrounded the office building where Nichols was holding hearings; at another meeting, Hells Angels crowded in to protest new motorcycle rules.[5] Later, when local regulators banned a toxic solvent, the dry cleaners who'd been using it came by the busload to protest. The state required special systems to capture the vapors wafting from filling stations. And Nichols deployed California's power as the only state allowed to impose its own rules on vehicles to demand one improvement after another to engines and fuel. Because carmakers didn't want to sell different models in different parts of the country, the changes she muscled through brought benefits all across America.

And slowly but surely, L.A.'s air began to clear. Well into the 1980s, parents still forbade kids from playing outside on the worst days. But as the steady improvement continued, those days grew rarer. After decades of tightening the lid on factories, power plants, and other big emitters, L.A.'s local pollution body, the South Coast Air Quality Management District, now regulates on a level more granular than just about any environmental agency in the nation. Its inspectors check store shelves for banned paints, and even nail polish removers must not exceed limits on chemicals—known as volatile organic compounds, or VOCs—that contribute to smog.

After a stint at the EPA, the agency she'd begun her career by suing, Mary Nichols returned in 2007 to lead California's Air Resources Board once again. Among the polluters in her sights this time were the million or so trucks on California's roads, collectively the state's single biggest source of diesel fumes.[6] She turned a local crackdown on their emissions into a statewide mandate to clean up. Most truck rules apply only to new vehicles, so health benefits take decades to materialize. But Nichols insisted those already on the road were covered too. The rule also applied to diesel school buses, whose fumes are breathed most intensively by the kids piled inside.

There are those who revile Mary Nichols, branding her air board an out-of-control destroyer of jobs. Many, even in the industries she's wrangled with, evince a grudging respect, and say she has always listened to their concerns, ready to meet halfway on practicalities if not big-picture goals. Environmentalists, too, have mixed feelings about Nichols. Some cast her as a congenital compromiser, too willing to water down rules. Others hail her as a savior, a "rock star" of regulation,[7] "the Thomas Edison of environmentalism"[8] who landed on *Time*'s list of the world's 100 most influential people. As with opinions of an ambitious young woman who'd been just behind her at Yale Law, Hillary Rodham, views on Nichols sometimes say as much about those who hold them as they do about their subject.

What is not debatable, though, is that L.A. today is a very different place than it was when Mary Nichols got her first look at it. I hear the same observation from just about all the longtime Angelenos I meet: Years ago, heavy smog meant you could hardly see the mountains. Today, their craggy beauty is the backdrop to everyday life. Now, ozone peaks at just a third of long-ago levels[9]—although it still exceeds legal limits—and L.A. meets federal requirements for several other pollutants. That the achievement has come as the region's population, its economy, and the number of miles its residents drive have soared makes it all the more impressive.

And yet. For all its good news, Los Angeles still struggles with a very real pollution problem. You can't see or smell it the way you used to, and most Angelenos don't give their air much thought anymore. But that doesn't mean it's not killing them. This is L.A.'s paradox: Its assault on smog is sometimes described, rightfully, as one of history's great environmental achievements. But this city's air remains among the country's worst. Greater L.A. consistently takes first place in the American Lung Association's ozone rankings, and makes the top five for particle pollution.

Fundamentally, that is the result of 17 million people and 11 million cars—plus the ship and truck traffic drawn by the nation's two biggest ports—crowded into a basin surrounded by those gorgeous mountains. This region's regulation is the country's most aggressive because it has to be. And despite cars' dramatic cleanup, the sheer number of them means Angelenos will probably never breathe truly healthful air until they run on electricity, or another zero-emissions technology.

What's more, as L.A. has notched up its successes, science has pushed the finish line further away, with finding after finding showing

pollution levels that would have seemed miraculously low to the gasp-
ing generations of years past are still blighting lives. And as is true in
so much of the world, while dirty air affects everyone here, some pay
a heavier price than others.

* * *

This visit to L.A. is my first in years, and like many outsiders, I find the
dry, sprawling landscape a bit alien. What strikes me most, though, is
the scale of the place. L.A.'s traffic and its freeways are famous, so I'm
not surprised to find myself sitting in bumper-to-bumper tie-ups on
eight-lane roads lined with palm trees. What I didn't anticipate was
just how vast the distances are. With my internal compass set by compact New York and slightly more spread-out London, I'm shocked by
the hundreds of miles I put on my rental car in just a few days.

The other thing that takes me aback is the very visible presence of
the oil industry. When I arrive a little early for an interview in one
pretty neighborhood, I follow signs to its wide white beach, and am
surprised to see the huge tanks of a refinery towering above the sand.
Oil pumps and derricks, sometimes camouflaged but often in plain
sight, sit next to homes, in McDonald's and Starbucks parking lots,
and just beside Beverly Hills High School. "They ruined a perfectly
good oil field by building a city on top of it," a lawyer for the industry
once complained.[10]

There's one name I hear again and again, whenever I talk to anyone about L.A.'s pollution: Jesse Marquez. Marquez is a tireless activist who's spent decades pummeling government and industry with
demands to clean up his neighborhood, Wilmington, on the southern
edge of the city. So, early on Super Bowl Sunday, I'm standing at his
front door, roses and orange trees blooming in the yard, as a pack of
tiny dogs yap wildly inside. He wanted to be sure we'd have plenty
of time to talk before the game, so he's invited me to come at 9:00,
although he got in late last night from a conference in Baltimore.

Marquez is a beefy man with a round face and a neatly trimmed
moustache; he's wearing flip-flops, jeans, and a black Jimi Hendrix
T-shirt, his hair pulled back into a tight ponytail that hangs down his

back. When he comes to the door, he's cradling a four-pound teacup Chihuahua in his arms, and several others are nipping at his ankles.

The verbal bombardment begins immediately. Over the next five hours, stroking the miniature dog in his lap, Marquez, who is in his sixties, will cover topics from Aztec history and the Irish potato famine's effects on Mexican politics to his glory days as a high school cross-country star. The subject that animates him most, though, is the air. Wilmington sits atop a major oil field, so drilling sites are scattered among the houses. There are three big refineries in the area, and a fourth just on its edge.

Jesse Marquez's unlucky enclave is also a major freight hub. The nation's largest container port, the Port of Los Angeles, occupies a big chunk of the neighborhood, and the second biggest, the Port of Long Beach, is right next door. Together, the two handle more than a third of the container shipments coming into and out of the United States.[11] Everyone knows about Southern California's traffic, but for those unschooled in the intricacies of air pollution, the role of the freight industry may be more surprising. Its invisible truth is this: In their journeys from factories to our front doors, the things we buy leave a filthy trail behind them. So Marquez and his neighbors, in the shadow of those two huge ports, breathe some of L.A.'s worst air.

In the parlance of experts, Los Angeles suffers from a regional air problem that plagues everyone, and also has hotspots where things are even worse. It's no surprise that Wilmington, a notorious hotspot, is poor and overwhelmingly Latino; others are largely African American. "Environmental justice" is the phrase often used for the struggle to ease that unfair burden, and it is the cause that drives Jesse Marquez.

Worried for his neighbors' well-being, he founded the Coalition for a Safe Environment in 2001. "I knew nothing about ports, international trade, nothing about how cranes operate, nothing about container ships, oil tanker ships, nothing about diesel trucks," he tells me. But the port's management was talking about building a wall at the edge of its land, and he and others were wondering what they wanted to do behind it.

It turned out plans for a big expansion were in the works. "We all said 'No.' I just yelled out 'Hell, no! Over my dead body.'" At churches and schools, wherever he could find a few people gathered together, he passed out flyers, spreading the word about upcoming meetings and explaining how profoundly bad air was affecting Wilmington's health. "We stood on corners and interviewed people," Marquez recalls. Three-quarters said someone in their family had breathing troubles.

Since then, it's been an all-consuming series of battles, in the courts, the media, and government offices, against one port expansion plan after another, against proposals to widen a freeway and build a rail hub, in favor of a new park. Marquez's days are filled with port commission and City Council hearings, and he's become an amateur litigator, working late into the night on voluminous briefs. "I'll prepare my 30-, 40-page written public comments with 3,000 pages of attachments." Once, "I appeared before City Council with my little helper elves, carrying these thousands of pages of documents. Boom, we delivered them. That becomes part of the administrative record." Of his opponents, Marquez boasts, "They never dreamed in their wildest dreams that a homeboy from the east side of Wilmington would have an IQ of more than zero, and be able to challenge them on every area of their expertise."

Marquez steps out of the living room for a moment, and comes back waving an old peanut butter jar that makes his point more viscerally than the flood of words and paper. It's filled with fine black powder, collected from the exhaust of a cargo ship. "Here's what we're breathing every day. Our lungs are the filters." He has invited a few neighbors and fellow activists to join us. While Marquez putters around the house, tidying and unpacking, then retreating for a shower, his guests wait patiently, sitting down one by one to tell me their stories. There is sadness in their words, an acute understanding of the suffering that blights this community. I hear something else, too: a sense of betrayal, by those whose noxious fumes cause all that hurt.

Ricardo Pulido has done everything he can think of to protect his family from the air, but he knows it's not enough. The Pulidos live in Carson, just next to Wilmington, surrounded by refineries. Years ago, he and his wife invested in double-pane windows, and on bad days they pull them tight to repel the stench—"real acidy, real bad smell." Hunkered down indoors, the whole family feels queasy, and they break out in itchy rashes. Heads ache, draining energy. "Dad," the kids complain, "I don't feel like doing anything." Pulido doesn't let them play outside on such days, and he's careful about his own activity too. "Normally, I walk every day, I see what's going on, I go to the gym once in a while. All that stuff is cut back."

Pulido has fought alongside Jesse Marquez for years. He's sweet and soft-spoken, with big, bushy eyebrows, a salt-and-pepper moustache, and dark blue baseball cap. But there's no hiding the anger beneath the surface. The refinery executives and port bosses, "they don't live around here. I always tell them, 'Where do you live?' 'Up in Topanga, down in Orange County.'" They make money in Wilmington, pollute in Wilmington, and leave it behind every night. Pulido enjoys talking to his neighbors, and sometimes stops by clinics to chat with parents who speak little English and have sought medical help only as a last, expensive resort. They blame their kids' nosebleeds and coughing on the atrocious air, but don't always share his sense of urgency. "They're so used to it," he tells me. "They think no one's going to change it, and they believe no one cares."

Sofia Carillo, too, sees the human cost. An elegant woman in a pink blouse, dark pinstripe trousers, and smart hoop earrings, she came to L.A. from Mexico in 1988. She was with Marquez in Baltimore, and she, too, got home late last night. This morning, she woke feeling like she always does in Wilmington: her voice was hoarse, her throat sore, and her head throbbing. "Every single day is the same," she says. "When I go to visit to my daughter in Austin, I feel good, excellent." As soon as she gets home, the symptoms roar back. Her doctor warns she is risking her life by staying here, but this neighborhood, for all its troubles, is her home, and she doesn't want to say goodbye. "I have my job, I have my husband, I have my community," she tells me.

Wilmington's suffering sometimes breaks Carillo's heart. A little girl she knows, only five, can't go outside without risking an asthma attack. So she sits at home and watches through a window as her friends play.

Carillo has been at port meetings where security issues were discussed at length, but in her view, the real threat doesn't come from violent outsiders. "The terrorists are inside. The refineries, the ports, the roads—these are the terrorists."

* * *

I've come to Jesse Marquez's house not just to talk, but because he has promised to show me around the neighborhood. It's getting toward lunchtime, so we start at his favorite Mexican takeout place. Within a few minutes, he's back behind the wheel, munching on a chile relleno burrito and drinking sweet tamarind juice through a straw. While he drives, I watch the dystopic landscape pass by. Our first stop is the ConocoPhillips refinery, a vast, twisting block of metal pipes that snake up, across, and around one another, beside a field of giant tanks. At the edge of the complex, right outside a low wall, sits a little house with a small yard. Out back, clothes dry on a line, and atop the garage is an air quality monitor Marquez has set up. The house is on a narrow strip, just a block wide, sandwiched between the refinery and the busy 110 freeway. A community college is a stone's throw away. Sometimes, Marquez tells me, it has to close because flaring at the refinery makes the air unbreathable. Next, we head toward Valero Energy's refinery, and then past one run by a company called Tesoro, both huge, with the same big tanks and twisting nests of pipes.

Along the way, I see how the movement of freight also shapes the neighborhood's topography. This place, and others like it, form the backbone of America's consumer economy. Shipping containers are everywhere, stacked five and six high and packed with the stuff of everyday life, from sofas and sneakers to car parts and toys. It's not just the two vast ports themselves, spread over thousands of acres, crisscrossed by their own internal roads and rail lines. Everywhere, we pass huge storage yards and truck and train depots, so many that I quickly lose count. They're all "diesel magnets," hubs for a never-ending parade of trucks that trail clouds of toxic exhaust as they trundle through Wilmington's streets.

The scale is stunning, and there are small houses and neighborhoods dotted among the industrial spaces. One row of trim homes, orange trees in their front lawns, sits across the street from a storage yard crowded with containers. Blight like this stretches along a freight corridor running inland from the ports. Heading east from L.A., freeways are clogged with trucks hauling just-off-the-boat containers emblazoned with the names of big shipping companies, destined for Targets and Walmarts across the country.

Many will stop first in the towns of the Inland Empire, which has grown into a distribution hub crammed with mega-warehouses: tiny Mira Loma—hot, flat, and dry—endures 15,000 truck visits a day.[12] Like most Americans, I rarely think about places like this when I buy the clothes or electronics they bring into my life, and looking out at all those containers, it is jarring to wonder what's inside. A pair of leggings I might, on impulse, pick up for Anna a few weeks from now? The new battery my phone is starting to need?

The system's tentacles, of course, reach far beyond California. Freight is the essential interconnector of a globalized world. And wherever it touches, it brings illness and death. Oceangoing cargo vessels like those docked at the ports here are among the worst polluters. Their engines—some as big as five stories high—run on the gloopy residue left over after refineries extract diesel and gasoline from crude oil. Known as bunker fuel, it sops up contaminants and is widely regarded as one of the dirtiest energy sources. It's viscous and heavy, so dense it has to be heated before it can flow through pipes. Bunker fuel contains up to 1,800 times more sulfur than would be allowed in a vehicle on land. The human consequences are severe, particularly since coasts are often densely populated. In Europe, for example, pollution from cargo ships is responsible for an estimated 50,000 deaths each year.[13] The climate suffers, too: shipping's carbon footprint is more than 2 percent of the world's total, larger than the annual emissions of Britain or Germany.

The good news is that Jesse Marquez's years of fighting have brought results. There have been changes at the ports that dominate his neighborhood, many springing from an epic battle over plans to expand a Port of L.A. terminal used by a company called China Shipping. The proposal—for a facility big enough to hold more than 9,000 containers, expected to draw a million truck visits a year[14]—became a proxy for the community's frustration with both ports' unfettered growth. So Marquez and other activists banded together with a determined team of lawyers to fight it.

In a standoff that dragged on for seven years, they used one legal lever after another to block every development project the port proposed. No new docks, no terminals: all construction would cease until the two sides could forge a truce. In a competitive industry, the lull was painful and costly for the port. Eventually, an appeals court ruled for the environmentalists, and harbor officials, desperate to resume growth, ponied up $50 million and promised to clean up. The Port of Long Beach, which had been watching nervously from the sidelines, signed on to a joint Clean Air Action Plan, eager to prevent its own dreams of expansion from being thwarted by a similar fight.

Mary Nichols's agency later turned many of the changes the deal ushered in into statewide mandates, so ports from San Diego to Oakland had no choice but to change. Much of it could have been done years earlier, of course. Cargo vessels need electricity while docked, and usually run their auxiliary engines to generate it. So just one visit by a big ship fouls the air as much as a diesel truck circling the world almost three times.[15] But if a port offers plug-in power, captains can turn their engines off. The navy had been doing it for years, but Los Angeles became the first to give commercial container vessels the option. Now, it bans idling for a gradually rising percentage of ships. Few competitors have followed suit, so in most harbors, moored ships still spew toxic exhaust.

Other changes were even easier: For ships unable to plug in, a cap can snap onto the smokestack and filter out more than 95 percent of pollution.[16] Incentive payments have persuaded shippers to cut speeds as they approach land, which halves pollution.[17] And the ports' big cranes have begun casting aside diesel to use electricity instead.

Nichols's agency went further, too, requiring ships to use lower-sulfur fuel as they approach the coast. Questions of jurisdiction—who makes the rules for vessels that circle the globe?—have long stymied efforts to clean up international cargo. So the new rule was a stretch, but courts backed Nichols up. Later, the prospect of relief came to other states, too; an international agreement will eventually require cleaner fuels to be used along all of America's coastline, not just California's. New limits will restrict sulfur to a twenty-seventh of previous levels—a clear hint of how high it has been for so long—and are expected to save 31,000 lives every year.[18]

So while it may be true, as the L.A. and Long Beach ports like to say, that their cleanup has led the industry, the reality is that this is a very low bar. Activists believe the ports were only doing what they should have done long ago: minimizing the illness and death their business causes. Nonetheless, it's clear the changes here have brought real benefits. As the new rules kicked in, diesel particulate levels in greater L.A. plummeted 70 percent over seven years. As a consequence, cancer risk from airborne toxins fell by almost as much.[19]

A few days before Marquez drove me around his neighborhood, I got a look at the Port of L.A. from the inside, on a management-sponsored boat ride. Aboard a little touring ship with a snack bar and a few dozen other passengers, I spent most of the time craning my neck to gaze up at the hulking vessels. Both the ships and the containers they carry are supersizing to keep up with demand for imported goods—the proportions of the average new cargo vessel have tripled in a decade.[20] Cruising along for an hour or so, we covered just a few of the port's 43 miles of waterfront. An announcer narrated over loudspeakers: One ship was importing steel, another bringing in blades for wind turbines, and a huge machine stood ready to shred scrap metal into tiny shards, for recycling overseas. Fields of oil tanks sat by the docks, and trucks rumbled overhead on a long bridge.

With further expansion on the way, and cargo expected to triple in coming decades,[21] it's clear the ports' cleanup must go further if improvements are not to be swamped by increasing volume. But now, harbor officials warn, intense competition and falling profit margins make another generation of changes harder to manage. For, of course, the stakes here are measured not just in pollution readings and cancer rates, but in dollars and jobs. Nearly $1.25 billion worth of goods pass through the two ports every day,[22] and the cargo industry underpins a big chunk of Southern California's economy. Now, the ports are scrambling to hold on to market share as the expansion of the Panama Canal gives Asian manufacturers the option of bypassing the West Coast to send goods directly to the eastern United States.

From Jesse Marquez's perspective, the shippers' and truckers' and oil refiners' claims that making them do things differently would hurt the region's economy ring hollow. The evidence shows protecting health nearly always saves money, if not for the polluters themselves, then for society more broadly. Indeed, one study found the health benefits of meeting Clean Air Act standards for particle pollution would save the L.A. region more than $20 billion a year.[23]

Some entrepreneurs are ready to embrace a better future. At the Port of Long Beach, a new kind of terminal is rising: the highly automated, partly solar-powered Middle Harbor is expected to generate just half the pollution of a traditional terminal, replacing trucks with trains and diesel cranes and tractors with battery-powered equipment. It's been hailed as a model for the future, proof that moving goods doesn't have to foul the air.

There's talk now, too, of electric trucks and cleaner shipping fuels, of overhead wires to power dockside machinery, and a trajectory toward zero emissions. It's not just communities like Jesse Marquez's that stand to benefit. Together, the ports of L.A. and Long Beach are Southern California's single biggest source of air pollution,[24] so while those living close to the freight hubs suffer most acutely, this industry's footprint touches everyone. There's a long way still to go, and progress could be so much faster. But amid all the arguing and litigation, this much is clear: The work of activists like Jesse Marquez and regulators like Mary Nichols has dramatically improved the air millions of people breathe.

* * *

Now, California is looking to the future. Nichols knows the cleanup she has helped bring about is not yet enough. Even when the skies look clear, she tells me, pollution exacts a price measurable in lost workdays, hospital visits, illness, and death. So she's still prodding truckers, shippers, and carmakers for the next measure of change. Of course, opponents of such progress are pushing hard to halt it, if not reverse it. Across the country in Washington, Republicans hope to strip California of the power it was given as the only state already regulating auto emissions when the Clean Air Act passed in 1970. So Nichols, and her colleagues in Sacramento, are on the front lines of a battle to prevent this state's air, and the nation's, from getting dirtier.

Another challenge has emerged too. As wildfires grow more frequent and intense in the western United States and Canada, their smoke has become a regular presence, and a real health threat. It's thick with toxins and particulates, and it blankets cities like Seattle and San Francisco as well as L.A., so bad sometimes it grounds flights, and forces authorities to distribute masks, and warn even healthy adults to avoid going outdoors. I experienced the smoke myself during a particularly bad fire season (although they seem to grow worse every year). Hiking in the Northern Californian mountains, my family and I gazed at views that would ordinarily be stunning, but were instead shrouded by thick haze. It reminded me of nothing so much as my visit to Beijing. And of course, the consequences—illness and, almost certainly, deaths—are far worse than just a few ruined snapshots.

Those wildfires, stoked by the heat and drought that global warming has visited on the West, are a reminder of what's now at stake. In addition to making air more breathable, Nichols's air board is charged with implementing California's ambitious climate agenda, aiming to turn a lofty goal of halving the state's oil consumption by 2030 into reality. Here, the two priorities—healthful air and a stable climate—are deeply intertwined, if not inseparable.

Mary Nichols is in her seventies now, and she sometimes invokes her grandchildren when she describes the future she envisions, one in which the state's air grows ever cleaner, electric cars replace their gas-powered predecessors, and carbon emissions fall while the economy grows. "Zero emissions" are words I hear surprisingly often in Los Angeles. In most places, the phrase signals a somewhat utopian vision of the future, or at least a distant one, and it usually falls from the lips, or the wonky policy papers, of committed environmentalists. Here, even port bosses, and a nose-to-the-grindstone bureaucrat I meet at the regional air agency, toss the phrase around. From freight equipment to cars, zero-emissions technology is talked about in California as if it is the necessary and inevitable next step, not some unachievable green fantasy.

The push for electric cars has been aggressive: The state offers generous incentives to those who buy them, and officials hope to get 5 million onto roads by 2030. Californians buy about half the country's plug-in cars, and it's one of the few places I've been where they are a common sight. An important part of the state's strategy is to help those of modest means buy clean vehicles—without that aid, they're likelier than the better-off to drive dirty old clunkers. One Californian, struggling to make ends meet, told me such subsidies and loans had been a lifeline for her. "If I hadn't bought an electric car, I couldn't have bought a vehicle that was capable of getting to work and back," she said.[25] If others follow California's lead, electric vehicles may someday soon shed their image as just another toy for the rich.

Los Angeles has come so far since the days of its neon sunsets that it's easy to believe its air problems are behind it. "One of the more frustrating things with my job is to convince those people who have been around a long time that we still have a problem," says Joe Lyou, a trim, tightly coiled man whose intensity is apparent the minute we sit down with iced coffees at a café near his home in El Segundo, a half hour from downtown. Lyou, the head of California's Coalition for Clean Air, is a believer in the power of government, of smart policy making—the step-by-step progress of Mary Nichols and others like her—to make lives better. He wants L.A. to be a cleaner, healthier city, and he knows it can be done, because he's seen it happen time and again. "It really is incredible how much progress we've made," he says. "It works, if you plug away at it."

Some time ago, Lyou tells me, an engineering expert was taking measurements at L.A.'s airport, trying out a new technique for tracing pollution particles' travels to pinpoint their sources. The scientist was puzzled by results that showed a plume of sulfur oxide originating a couple of miles offshore. "I said, 'That's the Chevron marine terminal.

Those are the ships that are parked off the coast'" waiting to haul fuel, Lyou recalls. The professor was surprised; he hadn't known about the facility. A few years later, he was back again, and the pollution-generating spot on the sea had vanished. "Why is it gone?" Lyou asks me. "Because Mary and [the Air Resources Board] adopted a regulation, said, 'You've got to change the fuels in those ships.' So I took the study results, I sent them to Mary, I said, 'Here's some nice data. Here's something that shows that it actually works when you adopt those regulations. Real factual data based on independent monitoring by a guy who didn't even know the damn thing existed.'"

It's just that kind of measurable effect that has made Nichols's career so gratifying. The excitement has dulled since her first, long-ago legal victory, through the decades of refereeing—and joining—battles between truculent industries and dissatisfied environmentalists. Looking back, though, she can't think of any other career in which she could have helped deliver more important, more lasting change.

The benefits of that cleanup may not elicit the thrill of shifts like the ones Mary Nichols's father fought for, the social change that led to the advent of same-sex marriage or the election of an African American president. But it, too, is part of the story of modern America's progress. And it, too, is a story that remains unfinished. While the work grinds on, L.A.'s successes show the rest of us that the fight for cleaner air is one that can, with science and persistence and the will to keep on pushing, be won.

10

LIVE FROM THE "AIRPOCALYPSE"

China's Next Revolution

The light pours down, at the end of its voyage, nearly a hundred million miles from where it began. The material it is about to meet, its soon-to-be partner, has been on a journey too. Silicon is among earth's most abundant elements, present in sand and clay, flint and quartz. To be readied for this job, it had to be melted down, purified. Then invisibly tainted, tiny imbalances injected into its otherwise perfect crystals. The silicon sits in layers, and while one is seeded with extra electrons, the other gets gaps where electrons are missing. That difference creates an extraordinary potential. A potential that needs just a little jolt from outside to be realized, to become something greater: power. The jolt comes from the sky, from the sun, and it starts the electrons flowing. Rushing to the border separating positive from negative, they hit tiny fingers of silver, then flow into a circuit. Electricity. Solar. Ready to do the work that keeps the world running, to light the lights, charge the batteries and the phones and the cars. Energy from a star, captured by materials from the earth. Beaming down to us from above. There for the taking.

"Sensitive" is the word used in China for the things its leaders deem unsayable, the issues from which those who want to avoid trouble must avert their gaze. Democracy, or a lack thereof, is sensitive. Making money, in a Communist nation full of energetic capitalists, is not. Air pollution used to be sensitive, until it grew impossible to deny, and the government finally decided to do something about it. These days, the media report openly on the filth that hangs over cities, companies woo employees with expensive office filtration systems, and ordinary

215

Chinese are free to complain about pollution on social media. Strictly speaking, the subject is no longer sensitive. But those whose views are a bit too pointed, or who speak through too powerful a megaphone, may still find themselves on territory touchier than they had imagined.

That's what happened to Chai Jing. Even without subtitles or any knowledge of Chinese, it's easy to see what made her a TV star. Young, smart, appealing, but not intimidating, she's the definition of telegenic, a serious reporter who also comes across as a regular person. With the ambition and savvy to navigate a journalistic career in a nation where press freedom is a distant dream but some issues are open to substantive coverage, she climbed from childhood in a northern coal-mining town to a prized job as investigative reporter and host on China Central Television, the main state broadcaster. She made her name with coverage of the SARS epidemic, and later, when she had become one of China's most popular television personalities, authored a best-selling memoir. Even then, her work was harder-hitting than an outsider might imagine possible. While direct criticism of the government is dangerous, she took on issues as varied, and provocative, as the ethics of the death penalty and attitudes toward homosexuality.[1] In the process, Chai became something of a hero to progressive Chinese hungry for an independent voice.

She left CCTV in 2014, and it was after that, when she was no longer hemmed in by the constraints of a privileged post, that she started on the project whose impact would be greater than anything she had done before, indeed greater than most journalists, even the most successful, achieve in the course of a career. She funded the $160,000 documentary herself, with earnings from her book. The finished product, *Under the Dome*, is at once engaging and terrifying, easily the most powerful work I've seen on air pollution. In faded jeans and a plain white top, Chai strolls across a small stage, telling a rapt audience that she discovered she was pregnant just after an awful "airpocalypse" in January 2013. She laughs shyly as the big screen behind her shows the ultrasound image in which she first saw her daughter, then describes hearing the child's heartbeat. But the story takes a dark turn. Joy turned to terror when Chai learned the baby had a benign tumor that would have to be removed right after birth. The big screen shows a teddy bear Chai clung to while her daughter was in surgery, and she says she held the unconscious infant's hand against her face afterward. "I called her name until she opened her eyes and looked at me."

Chai had never worried much about pollution, but the responsibilities of motherhood quickly changed her perspective. "This," she says, "is when you feel threatened." Well aware now of the damage Beijing's foul air does to young lungs, she says she lets her daughter outside only when air quality is good. That means the little girl stays indoors half the year. Sometimes, her mother spots her gazing out a window. "Sooner or later, she will ask me, 'Mom, why do you keep me at home? What is going on outside? Can it hurt me?'" Chai tells the audience. Her work on the documentary, she explains, has been about finding answers to those yet-unspoken questions, and some others too.

In the course of the film's 104 minutes, mixing clear, straightforward narration with video clips of her reporting, Chai lays out the dangers of the tiny PM2.5 particles that plague China, explains where they come from, demonstrates government's failure to confront the crisis, and describes how other countries have done better. She displays a filter that was midnight black after she carried it in a backpack sampler for 24 hours, and says she and a scientist she engaged were stunned by the results when he tested the soot on the paper. It contained 15 different carcinogens, including highly toxic benzo(a)pyrene, at levels 14 times China's legal limit. "I am in the center of an international metropolis. Looking around, I cannot see any factories or chimneys," she says. "How can the amount of benzo(a)pyrene be 14 times the standard?"

The jagged, irregular image of a pollution particle, captured through a microscope, fills the screen, and Chai explains it is structured like a chain, so that if you could unfold it, you'd see an area as big as a basketball court. "That is why it can absorb carcinogens and heavy metals," she says. In an animated sequence, the particles are depicted as cartoon villains, skiing down the lungs' passageways and defeating the body's white-armored defenders. "Tremble, human!" a particle intones. "I am there! Always!"

Chai acknowledges the complexity of her story, too, since as well as tainting the air, China's rapid economic growth and the urbanization that has accompanied it have also raised living standards and given her the chance to build a big-city life more fulfilling than the one she expected to live in her provincial hometown. She grills industry executives and bureaucrats, and ultimately lays blame on a government that for years allowed petrochemical companies to write fuel standards, failed to punish factories that flout pollution laws, and turned a blind eye to the widespread use of fake papers to claim dirty trucks and cars meet emission standards. She visits California, and marvels at the achievements of Mary Nichols's Air Resources Board. Switching to English, Chai questions a trucker who's been fined for failing to install a filtering system. And she explains how the rule of law—the system of courts, of even-handed enforcement essential to the functioning of advanced democracies but nonexistent in China—ensures that those who violate America's pollution rules are, for the most part, caught and punished, while the opposite is true in her country. In retrospect, homing in on a flaw so fundamental to the way China's government works may have been risky.

The public's response to *Under the Dome* was overwhelming. Within days of its release online, in February 2015, more than 200 million Chinese had watched it, an audience three times the population of Britain or France. "This really went viral to a level that nobody had envisioned," Ma Jun, one of China's preeminent environmentalists, told me. The government appeared to be on board, and, indeed, it seemed Chai had sought—and obtained, at least informally—official approval for the film; she told *People's Daily*, the Communist Party's newspaper, that she'd shown the script to two different government entities and gotten feedback and suggestions.[2] The environment minister compared her documentary to Rachel Carson's 1962 book *Silent Spring*, which, in exposing the ravages of the pesticide DDT, inspired the modern American environmental movement. It looked like China's leaders were planning to harness the outrage *Under the Dome* stirred up to generate political momentum for tough new rules they were preparing to impose on polluters.

But then, something changed. A week after it was posted online, *Under the Dome* vanished from domestic websites, removed at censors' behest. It's impossible to know what caused the about-face. It seems likely there was disagreement within government, a hard-line faction gaining the upper hand over more open-minded colleagues. The film's astonishing popularity clearly caught officials off guard, and they may have feared Chai had unleashed something they couldn't control, something that, if left unchecked, might prove potent enough to undermine their precious legitimacy, their very ability to hold on to power. Few things are as frightening to China's leaders as the prospect of social turmoil. So the anger of hundreds of millions of people united around one issue must have felt dangerously destabilizing, a potential that needed to be quashed, or defused. Official decision making in China takes place behind closed doors, though, so those whose lives are shaped by leaders' edicts can only speculate on how and why they are arrived at.

I hear a great deal of such guesswork in Beijing. Chai Jing was simply too outspoken, the documentary too vigorous an exercise of free expression, to be allowed, one expert tells me. The government was fine with the film's message, another says. The trouble was that it had gotten bigger than the issue it addressed, and Chai herself had become the story. Actually, another asserts, even that would have been permissible: the problem was the timing. Chai released her film just before the "two sessions," the annual gatherings of the National People's Congress and the Chinese People's Political Consultative Conference, a period during which authorities prefer that no other news interfere with the political theater of thousands of delegates coming together in the Great Hall of the People on Tiananmen Square. Authorities simply found it unacceptable, at such a moment, for a journalist to roil the waters of public discourse the way Chai did.

She surely understood the significance of the moment she chose, a moment guaranteed to amplify her film's impact. Perhaps she even planned it with allies inside government to help them push air quality up the meetings' agenda, to put public opinion on their side in a hidden power struggle over how aggressively to confront big polluters, how quickly to shift China's economy away from its dependence on the heavy industries that poison its air. One expert I meet—he sounds a bit envious of Chai's fame—even suggests she released *Under the Dome* when she did because she wanted it to be banned, so she would gain celebrity in the West and a green card to live in the United States (elsewhere, I see speculation that she holds American citizenship, which would make permanent residency superfluous).

For a censored video, *Under the Dome* is surprisingly accessible. One man I meet in Beijing pulls it up on his phone as we chat; a café is showing it on a screen visible to passersby. While China blocks sites like YouTube and Google with a vast system of online censorship known as "the Great Firewall," many download software that connects to an overseas server so they can view whatever they want. In fact, I didn't find time for *Under the Dome* at home, so I used just such a virtual private network, or VPN, to watch it on my first night in Beijing. Trying to be surreptitious, I listened through earphones so Chai's voice wouldn't be audible outside my hotel room, but that seems silly in retrospect, and I doubt anyone walking past would have cared even if they had overheard. As in other areas of Chinese life, what is sanctioned publicly and what happens behind closed doors are two very different things.

But while her work remains available to those with the wherewithal to seek it out, Chai Jing herself has dropped out of sight. Since the flurry of stories in the international press about the film's removal, she's barely been mentioned in any news article I or my Chinese-speaking research assistant can find, and is nowhere to be found on social media, Chinese or Western. The theories on what's become of her are even more confusing than the speculation I heard about why *Under the Dome* was blocked.

What is clear is that she herself has become a sensitive subject. Just inquiring after her whereabouts often seems to elicit anxiety, and when I put the question to the same set of informed guessers who speculated about the censorship decision, I get a mix of answers just as contradictory. She's in Beijing, I'm told, and is absolutely fine, enjoying a quiet life. She's gone to the United States, I hear later, although if that's true, I can't see why such a well-known media personality, fresh off the biggest success of her career, wouldn't have said a public word for years, or set up so much as a Twitter account. An environment expert I thought would be in the know professes ignorance, but believes she's safe and says he wishes her well. An American writer recalls hearing whispers that she's under house arrest, but that's as hard to verify as the other theories.

I can't stop wondering what's become of her, and I know it's a long shot, but I want to meet Chai Jing. Perhaps it's just the effect an attractive TV personality has on everyone who sees her on-screen, but I feel a connection to her. In part, that's because she's a fellow journalist. And of course, *Under the Dome* covered the same issue I've now spent years researching and writing about, albeit from a different perspective. Her reporting, while focused on China, has taken her to some of the same places I've visited. *Under the Dome* includes footage from London, where Chai stands in front of Battersea Power Station on the River Thames and explains how Britain moved from coal to natural gas after the Great Smog of 1952. She even interviewed Ed Avol, the Children's Health Study scientist I met at the University of Southern California; I started, in my cramped Beijing hotel room, when his face appeared on my laptop, subtitles translating his words into Chinese.

Of course, there's a crucial difference between Chai Jing and me. In the country I come from, and the one I live in now, people in my line of work might worry about a story's ability to draw clicks, a book's sales potential, but we don't have to wonder whether a subject's political implications might endanger us personally. In the United States and Britain, in any nation that protects free speech, hard-hitting journalism that confronts readers and viewers, even—no, especially—political leaders, with unpleasant truths is the path to professional advancement, to the front page and the top of the website, to promotions and prestigious awards. Where Chai Jing comes from—a nation that jails more journalists than almost any other, where a crackdown on the press has escalated in recent years[3]—such reporting can lead its practitioner to a far different fate.

Eventually, I have to accept that I am not going to meet her. Every inquiry, every email, has come up empty. Back home, I contact Amnesty International, the Committee to Protect Journalists, Reporters Without Borders. Chai is not on a list of jailed journalists, but beyond that, no one offers answers. Perhaps that's good news, and an academic I know assures me the grapevine would be alight with rumors if a celebrity like her was under house arrest. On the other hand, Reporters Without Borders' Asia director worries she's in trouble, perhaps detained in a hotel room, or some similar, informal prison. "In China, everything's possible," he tells me.[4]

This much, at least, is clear: A journalist who did her job well, a journalist I envied, when I watched her work, for cutting so powerfully to the heart of her subject, has either been silenced or has silenced herself. In the end, I'm left with just the power of Chai Jing's voice, of the extraordinary documentary that spoke to the fears and hopes of 200 million people. That, and an awareness that the question of just what is in the air their people breathe is still—despite the very real change Chai Jing's work helped spur—one that makes China's leaders uneasy. Still, after everything that's happened since *Under the Dome*, sensitive territory.

* * *

If there's one thing the world knows about air pollution, it's that China has a lot of it. But I haven't traveled here to rehash old news. I've come to Beijing to tell a different story, an optimistic one, the story of a country beginning, slowly but surely, to move in a new direction. A story about leaders finally taking action to protect their people's health, about the biggest investment in wind and solar energy the world has ever seen, about the planet's largest coal consumer scaling back its use of the foul fuel, and its carbon emissions, ahead of schedule. About air quality that, while still awful, is getting measurably better.

All those things are happening, and the hopeful narrative is real. But you have to squint hard through Beijing's filthy air to see it. Bleary-eyed after an overnight flight from London, I can barely make out a thing through the airport windows. Buildings just a few hundred yards away are visible only in outline; everything is shrouded behind a dark, heavy veil. In most of the pollution-plagued places I've been—Delhi, Krakow, the San Joaquin Valley—chance has had it that I arrived at relatively clear moments, and locals told me I would have seen far worse had my timing been different. In China, I've landed in the middle of, if not a full-blown airpocalypse, a typically disgusting Beijing winter. I have never seen air this bad, never had to breathe it for days on end, never had to contemplate the effect on my body of particulate levels that average six times higher than London's.[5] And there's no disguising the painful reality of its impact: The air kills more than a million Chinese every single year—a million and a half if the effects of household pollution are included.[6] Even in a country as big as this one, that is an astonishing toll.

After the shock of my arrival, the next day is clear and pleasant. But I'm told it's one of the year's best, and soon the thick, dark cloud of pollution is back, sitting heavily over this city of 22 million people, a gray place whose sprawling ugliness doesn't do justice to its vibrancy—a bustle and an energy that remind me of New York's. Rows of identical concrete apartment towers sit beside gleaming new office buildings and malls, many designed by celebrity architects with avant-garde sensibilities. Chai Jing's old employer, CCTV, has the most stunning, two leaning glass towers connected 40 stories above street level by a jagged, cantilevered V. It's a city where traffic inches along the half-dozen concentric ring roads that are its main arteries, and life proceeds on a more human scale along the hutongs, or alleyways, that haven't been cleared to make way for something new. And it's a city that's growing so fast, the subway map in my tourist's guidebook—the newest edition, less than three years old—is hopelessly out-of-date, missing so many new lines and extensions, it's all but useless.

That growth has come at a price: the pollution crisis China's people live with every day is the result of decades of unrelenting development. Soon I find myself checking the air quality reading online several times a day; depressingly, it seems stuck on red, "unhealthy." On the bright side, Beijing is the only place I've been where pollution masks are socially acceptable. Still, I suspect I look a little ridiculous in mine. It's a sort commonly worn by cyclists in London, black and somewhat Darth Vaderish. Although I emailed my facial measurements to customer service, as suggested, before ordering, it feels too small, squashing my nose painfully. An instructional video said air leaking out at the top or bottom indicates the seal isn't right and the mask will do little good. Over and over, I exhale hard to check it, repeating the procedure every time the mask slips, or I shift it in search of a comfortable position. Making matters worse, whenever I put the mask on, or adjust it, the Velcro fastener at the nape of my neck gets snagged on my scarf, my hat, my gloves. Plus, I've picked up a cold, and the pressure on my nose makes it drip beneath the mask.

That's the state of affairs as I trudge down the street behind Xiao Jin, a stylish twenty-something whose translation and reporting help are essential to my work here. (I'm withholding her full name to avoid causing her trouble; Xiao is a title often used for a young person.) We have emerged from the subway into an unexpected snowstorm— unexpected by me, anyway: in my focus on pollution forecasts, I have neglected to check the weather. Later, I see a reporter I follow on Twitter has called it "toxic snow."[7] The Beijing Meteorological Bureau sometimes warns residents to avoid touching snow, which can be dangerously dirty here.[8] Plenty of it blows into my face, and Xiao Jin's. The mask may not be doing much for my lungs, but at least it keeps those poison flakes out of my nose and mouth.[9]

* * *

Despite the still-atrocious state of this country's air, Ma Jun embraces the hopeful narrative about China's pollution. He is perhaps the country's foremost environmentalist, a modest-looking man, balding and slight, in a navy V-neck sweater and white shirt. Ma believes something remarkable is happening in China, and he can boil it down to a single word: transparency. Chai Jing found the limits of the government's tolerance for openness, but even so, air pollution is an issue on which leaders are uniquely accepting of free discussion.

Things have changed faster than Ma could have imagined. When Beijing hosted the 2008 Olympics, many Chinese, blissfully unaware of the health emergency unfolding around them, scoffed at foreign athletes who put on pollution masks. Even Chai Jing said in her documentary that she'd been blind to the crisis as recently as 2013. Until then, news reports generally referred to pollution episodes as fog, or bad weather, so dirty air wrought its damage without anyone taking much notice.

It was American diplomats who helped change that, albeit inadvertently. The U.S. Embassy in Beijing had been measuring pollution with a monitor on its roof and tweeting hourly readings, mainly for its own staff and resident foreigners. By 2011, anger had begun bubbling more widely. PM2.5 is the biggest hazard in the country's air, but China's government was omitting it from official pollution reports, so their data painted a rosy picture that appeared increasingly laughable. One analysis found that while the embassy readings showed pollution at unhealthy levels on 80 percent of days, government numbers indicated air was good or better 80 percent of the time.[10] Residents could see the heavy gray haze they were stewing in, but at the end of the year, the local environment bureau cheerfully reported the city had enjoyed 274 "blue sky days" in 2011.[11]

Ordinary people—and some big names with millions of online followers—took to social media. With VPN connections, they copied the American data from Twitter and shared it on Weibo, a popular Chinese site, sometimes side by side with the much lower numbers from their own government, making the official dishonesty plain to see. Once, when pollution spiked above the upper limit of the rating scale the embassy used, its normally sober terminology was replaced with a more colorful assessment: "crazy bad." The comment was apparently published accidentally, and the Americans quickly rephrased, deeming the day's pollution simply "beyond index," but not before Beijingers had captured and shared the more vivid words. China's leaders didn't take kindly to the embarrassment, and demanded that the embassy keep its readings under wraps. Like the pollution choking Beijing, though, the tweeting wouldn't relent.

And then, suddenly, everything changed. In retrospect, Ma Jun says, the social media uprising of 2011 was a watershed moment.

Government, it seemed, had heard the public outcry, and the years of stonewalling came to an abrupt end. Leaders announced they would begin measuring PM2.5 in cities across the country and releasing the data. Soon, Beijing had the first official monitoring station, and a handful of others quickly followed. By 2013, real-time readings were available from 74 big cities. "One year later, that number was increased to 190. And then another year later," it jumped again, to 380, Ma says. He clicks his mouse a few times, and on a big screen in the conference room where we're sitting, a map shows live pollution data from more than 2,000 monitoring stations in almost 400 cities.

In Ma's view, that made what came next all but inevitable, the power of information inexorably making itself felt. "People are not satisfied only to know which day to put on face masks. They want to solve the problem," he says. "The moment you decide to disclose that, to give people the truth, then there's no way for you to backtrack. It's one way, you have to move forward." And that—in his telling, anyway— was how China began to push back against its terrible air. In 2013, the government released an action plan that included some big changes. Among the most consequential was the decision to start reducing coal use in the regions around Beijing, Shanghai, and Guangzhou. China's coal consumption had tripled since the start of the century.[12] Before the turnaround, Ma says, it was expected to double from 2011 levels before eventually declining. Given how bad things were already, "that was totally unimaginable, and unacceptable. But that was the projection, that was the plan." Under the new policy, the heavy-industry belt around Beijing would have to use 80 million fewer tons of coal a year. "That was quite dramatic."

Besides transparency, Ma's other favorite word is data, and having gotten some of it, he wanted more. With air quality numbers for Chinese cities publicly available, the next thing people needed to know, he believed, was just where all that pollution was coming from. With a coalition of other groups, his Institute of Public and Environmental Affairs decided to push for the release of monitoring data on individual factories, so their neighbors would know when they were breaching pollution limits. The advocates thought it might take five years to wrench that information from its protectors, and began preparing to petition individual cities and regions for publication. Instead, it happened in one fell swoop, less than five months after the environmentalists had set out the goal. In mid-2013, the central government ordered every province to begin collecting and releasing those data at the start of the following year. "This represents real political will," Ma says. "Because when you allow people to know this, you cannot continue to protect these polluters."

That powerful faith in openness is remarkable in a man whose nation holds its secrets so closely. Ma believes so strongly in part because he has to. Unlike their counterparts in my own countries, China's citizens and watchdog groups can't take polluters, or the government, to court to demand laws be enforced (although a brave few have tried). Public pressure is the best tool available here. And Ma thinks it's working. His institute built an app that draws on all the new government data, allowing anyone to see, in real time, with a few taps of a cell phone screen, which steel mills, which cement factories, which power plants are exceeding legal pollution limits. With a pinch and a scroll on his phone, he pulls up another map and shows me the red and blue dots of real-time monitoring.

Not everyone was so excited about the new system. While the central government had decided to forge ahead with its anti-pollution program, some provincial governments were wary, reluctant to change. So Ma was nervous when he was called in to see the top environmental official in heavily industrial Shandong Province. To his surprise, the official began by praising the transparency push. "He said, 'I have the duty to try to control the pollution, but it's hard and we need people to join the efforts,'" Ma recalls.

Shandong's environment bureau set up a Weibo account, as did 17 of its cities and more than 100 counties, Ma says. Using Ma's app and the official data flowing through it, residents could soon post publicly whenever they saw a factory was violating emissions limits, and tag relevant enforcement agencies. Thanks to smartphones and social media, "for the first time, the local government can have some real, day-to-day interaction with the local people" on a pressing issue, he says. "This is a very rare opportunity."[13] I've never heard of such a sophisticated real-time reporting system in the United States or Europe. To encounter it in China is nothing short of extraordinary, yet another glimpse of the complexity and contradictions of a nation that, while among the world's most politically repressive, is also on the cutting edge of modernity.

It's clear even to a visitor that social media plays a huge role in Chinese life. Twitter and Facebook are blocked, their unfettered flow of information deemed unacceptable, but Weibo and WeChat are popular equivalents. Several people warn me it's hard to function in China without WeChat, so I download it before my trip; Dan does the same, so we can stay in touch while I'm away. To no one's surprise, Anna quickly hijacks his account, so my interviews with environmental activists are punctuated by the buzzing of my phone, as she logs on before school to send me dancing cartoon characters, chatty voice clips, and photos of herself.[14] Several of the people I'm seeing create WeChat groups ahead of our meetings to hash out logistics, and locals also use it to pay their bills, summon taxis, and send money to friends.

Ma Jun tells me social media has brought a more profound change, too, to the very nature of the relationship between China's leaders and its people. It's happened on air pollution, he says, because of the unusual degree of consensus on the issue. Everyone knows the dangers of dirty air must finally be confronted, and that has created a unique opening. "There's a chance to try a different way, almost like a different way of governance," he says. "We are seeing something very special."

* * *

Wu Bai takes a darker view. That's not his real name; a journalist who knows well what cannot be said publicly, he spoke to me on condition of anonymity. The Chinese government is under no obligation to respond to its citizens' demands, he says. The only way of forcing leaders to heed public concern is through elections, and without them, public opinion "doesn't count, it doesn't matter." So whatever China's leaders are doing about air pollution, they're doing it because they want to, for their own inscrutable reasons, not because anyone has demanded it. Those reasons, of course, may include a genuine recognition that it's a serious problem, sharpened by the knowledge that they themselves are affected, since Beijing is among China's most polluted cities. "Maybe they read something, they really think it's a bad thing," he says. "We're only guessing, because no one knows," except those on the inside.

I posit a notion I hear now and then, usually only half seriously, that problem solving can be swifter in a nation unbound by the messy constraints of democracy, that China, once it's put its mind to something, can move more effectively than a country obliged to heed voters' sometimes contradictory wishes. Wu dismisses the idea. "It's gambling," he says, "just waiting for the wise authoritarian government to have the good decision." The wheels of democracy may grind slowly, but when change comes in a free nation, it's more firmly rooted, he believes. And anyway, he points out, China's relatively swift moves on pollution follow years of miserable waiting.

The media aren't much help, says Wu, leaning back in his swivel chair and fiddling with his laptop as he talks. The need to tread carefully is deeply ingrained in China's journalists, and they know how to read the signals of what's allowed and what's not. "It's like an instinct," he explains. "Self-censorship is our training." Chai Jing was a rare exception, but nowadays reporters mostly follow where ordinary people have forged ahead. Social media makes it easy to see what's safe. If a given issue, a given viewpoint, is represented on Weibo or WeChat, the government has decided to allow it. Anything that vanishes shortly after it's posted is out of bounds, deleted by the censors who patrol Chinese cyberspace. When it comes to air quality, things can be complicated. An American academic who tracked Weibo comments on pollution drew this conclusion: Those focused on scientific data were unlikely to upset the censors, but politically tinged comments critical of the government were generally removed.[15] His research didn't touch on journalistic work, but that certainly sounds like the line Chai Jing crossed.

Many see Wu's election-centric view, his focus on democracy, as too narrow, failing to take in the complexities of China's political dynamic.

For centuries, this country's leaders have valued few things more than stability, and have moved to defuse any issue with the potential to threaten that equilibrium. Here again, social media has proved a useful tool, giving them a unique way to take the temperature of public opinion. It told them clearly—in 2011, and again in 2013, when an awful January pollution episode sparked a new wave of online activism and the term "airpocalypse" came into use, and yet again in the wake of *Under the Dome*—that anger over filthy air was reaching the boiling point. Pollution clearly held the potential to bring angry protesters onto the streets. "It could really destabilize society," Ma Jun told me. Avoiding that scenario has become a top priority.

* * *

It doesn't take an expert to spot the source of this country's pollution. While cars and trucks of course taint the air in Beijing and other big cities, and Chai Jing was right to identify poor fuel quality as a serious problem, the biggest culprit is coal. Pouring from the smokestacks of factories and power plants, and the chimneys of families who burn it for heat, its smoke is responsible for 40 percent of China's particle pollution. With its economy growing at breakneck speed, China's coal use quadrupled between 1990 and 2013;[16] it accounts for about half the world's total coal consumption. I see the evidence from one of the high-speed trains that zoom up and down the densely populated east coast, where one industrial plant after another belches its smoke into air already heavy with a dark fog. Unable to concentrate on the notes I'd meant to transcribe, I gaze out the window instead, transfixed by the sheer awfulness of the air.

Of course, this is not a one-dimensional story. Coal may be making China's people sick, but the growth it powered has lifted hundreds of millions from abject poverty in the space of a generation, surely one of humanity's greatest economic achievements. That growth forged an urban middle class whose lives would be unrecognizable to their peasant grandparents, and allowed young people like Chai Jing and my translator Xiao Jin to realize dreams their parents could never have imagined. But the price has been steep. China's water is tainted with heavy metals and agricultural runoff, contamination makes much of its soil unfit for farming, and entire villages are poisoned by toxins from mines and refineries. Now, economic security—a job, a home, enough to eat—is no longer sufficient. China's people are comfortable enough to want something more, and breathable air doesn't seem like too much to ask.

The government, too, has begun to reckon with the fallout of industrialization. A milestone in that reckoning came in 2014, with a vow from Premier Li Keqiang: "We will resolutely declare war against pollution as we declared war against poverty." Dirty air, he said, was nature's warning against "inefficient and blind development." It may certainly be true, as Wu Bai believes, that public pressure wasn't what prompted the declaration. But it was probably at least part of the equation. Air quality has provided political cover for some painful decisions that needed to be made for reasons having nothing to do with the environment. China's economy has been bedeviled by excessive production, particularly in the heaviest—and dirtiest—industries, like steel and cement. Kept afloat by endless credit and public subsidies, factories have been running at high gear for far longer than demand justified, creating mountains of debt that could trigger a major economic crisis. Leaders know they must dismantle this house of cards, but unleashing market forces isn't easy in a nation where state-run companies are big employers.

Air pollution has put the power of public opinion on reformers' side. Consistency is not the Chinese government's hallmark, and it has zigged and zagged in its effort to liberalize the economy. Leaders talk of reining in output, then rev it up again with stimulus measures, fearful of the consequences of truly scaling back. So their prosecution of the pollution war has been bumpy and lurching, and its progress often has a two-steps-forward, one-step-back quality. But it's clear this effort is serious. Officials are taking real-world steps like ordering factories and power plants to install modern air scrubbers and filters, tightening vehicle standards, upgrading gasoline and diesel quality, and demanding ships use cleaner fuels when cruising rivers and docked in ports.

Publishing policy papers in the capital, of course, is one thing. Turning them into reality across a nation as vast as this one is another, and enforcement of Beijing's edicts has always been troublesome in

China. Local officials may smile and nod at pronouncements from on high, but that doesn't mean they'll act on them. Over decades, provincial officials grew accustomed to being judged harshly by superiors if their bailiwicks failed to meet economic growth targets. In a rigid top-down system, such assessments can make or break careers, and while the evaluation criteria have been overhauled to include environmental protection, some functionaries have found old habits hard to break, and continue to push development at the expense of all else.

And the anti-pollution tactics can be heavy-handed, sometimes verging on absurd. Mayors who fail to deliver on air quality are called to Beijing to be reprimanded, the summonses announced publicly to intensify pressure. Officials order polluting plants shuttered to ensure clean air during high-profile events like international summits. In the northern city of Shijiazhuang, they closed factories for weeks one November and December so the area wouldn't breach its annual pollution limit.[17] Such drastic, last-minute measures may be meant to show seriousness, but in fact they demonstrate little more than poor planning.

Despite all that, the change under way is real. Leaders have built on the Beijing-area coal cap Ma Jun found so remarkable with a nation-wide limit, and are reducing the percentage of energy that comes from the fuel that, in addition to fouling China's air, accounts for a big chunk of its carbon emissions. Just as in manufacturing, excessive production and poor implementation of big-picture goals plague China's energy sector too. So new coal-fired plants are still being built, rubber-stamped by provincial officials who prefer dirty energy from their own backyard to clean power generated in someone else's. But many will never run at full capacity. And for the first time, significant numbers of coal plants already in the planning pipeline are being canceled. The numbers can be murky and hard to read, but they are edging downward, and—sooner than many had dared hope—experts now believe the peak years of Chinese coal burning are in the past.

* * *

The fruits of that progress aren't always easy to see. Tangshan is one of the battlegrounds on which China's war against pollution is being fought. An hour and a half's ride from the capital on another of those bullet trains, it's a hub of heavy industry—steel mills and cement plants and coal-fired power stations. The landscape is dystopic: rusting, hulking structures are surrounded by twisting pipes, big tanks, and—everywhere I look—smokestacks. I've never seen so many factories, so close to one another, spewing so much smoke. This is a different world from Beijing's modern shopping malls and sparkling subway. But it is this China that has made that other China possible, has provided the concrete, the steel, and the economic muscle to build it.

Between Tangshan's factories I see row upon row of tall, characterless apartment towers, 15 or 20 stories high, clustered together in groups of identical structures—a complex of beige buildings, another of white, one of red and yellow. That's a lot of people breathing the muck that passes for air here. Hebei, the province where Tangshan lies, is home to the top six worst-air cities in China.[18] It drifts elsewhere, too; the region's industry is a big contributor to Beijing's choking haze.

Chinese steel, of course, is a hot-button issue in both the countries I call home. This industry once brought steady paychecks, and the pride and social stability they engender—along with steelmaking's thick smoke—to Pittsburgh, Pennsylvania, and Youngstown, Ohio, to northern England and southern Wales. China's cheap production was blamed by Donald Trump for the collapse of employment in the deindustrialized midwestern states that propelled him to power, and by the forces that brought about Britain's wrenching decision to leave the European Union. It turns out to be a little touchy here too. When Xiao Jin bought our Tangshan train tickets, she indicated a destination one stop short of the city, worried someone might question why I'd wander so far off the usual tourist trail.

Wang Jing Bo has built his life on this disputed territory. Xiao Jin and I meet him next door to the Tangshan Guofeng Steel and Iron Factory, where he works with molten steel, purifying it, oxygenating it, tempering it. His wife runs a gritty little shop and restaurant, serving the drivers whose big flatbed trucks trundle by outside, laden with rolls and rods of steel. A small coal stove sits on the store's concrete floor, filling the air with smoke, and an assortment of items—flashlights, umbrellas, batteries—hang from a metal rack. Hard-boiled eggs are piled in a black crate on the floor.

Wang perches on a pink plastic stool beside a bank of listlessly blinking video gambling machines, and explains, while Xiao Jin and I sip from bottles of room-temperature coffee, that he has mixed feelings about the mill. The work is dangerous, the protective gear he must wear is heavy, and temperatures can soar above 120 degrees Fahrenheit. But the pay is good, and a bonus comes when the factory does well. He knows men who've lost their jobs when other factories have shut down or moved, but he says his own bosses run a good operation, making quality steel and following all the rules, so he's confident his job is safe. The industry's painful restructuring is necessary, he believes, and the best companies will survive. "It will be stronger and stronger rather than bigger and bigger."

Not everyone is so sanguine. A few miles from the shop, we meet another steelworker, a man with stylish glasses and spiky hair, wearing a green camouflage jacket over a gray hoodie. He doesn't want his name printed, perhaps because his words are sharper than Wang Jing Bo's. He works at another plant, and his salary has shrunk by 20 percent in five years. "Our lives are greatly and obviously affected," he tells us. He knows multiple forces are bearing down on his industry, and it makes no sense to produce steel no one wants to buy. But air quality rules are part of the picture too. He says his company spent nearly $150 million upgrading, in part to meet environmental requirements. "I have to say the renovated production lines are stunning," he concedes, and "this brings us more opportunities," like a new contract making steel for cars. And of course, he, too, must breathe the air here. A colleague commented recently that Tangshan was a good place to live, the pace of life comfortable, the sea nearby. "Except for the pollution," he says. "There are only a few blue skies all year." He hopes his son will find a better home someday.[19]

It's clear that powerful political and economic crosscurrents are at work in China's war on pollution. Nonetheless, the results so far are impressive. In nearly 200 Chinese cities, including Beijing, particulate levels fell by more than 16 percent in the year after Premier

Li's declaration.[20] They spiked again when steel production rose—the consequence was visible in the skies when I visited. But despite such backsliding, the gains are coming with remarkable speed. By 2018, four years after Li launched the pollution war, particle levels in Chinese cities had fallen by nearly a third. Those reductions, if they hold, will give every single Beijinger more than three additional years of life; in Shijiazhuang, even bigger improvements are expected to add more than five years to every life.[21] Such numbers are a powerful illustration of the benefits that come when leaders get serious about pollution.

Back in Beijing, I meet Tonny Xie at the Clean Air Alliance of China, a grouping of experts that advises the government. The language he uses sounds familiar, phrases like "attainment planning" and "emissions inventory." I heard them outside Los Angeles, where an air quality regulator named Elaine Chang told me about the years she'd spent working the levers of bureaucratic power to push Southern California's long struggle with smog a little further forward. It turns out Tonny Xie knows her. "She was our guest," he says with a smile, recalling two visits Chang made to share Los Angeles's pollution-fighting experience with Beijing.

China's struggle with dirty air is sometimes likened to what L.A. went through decades ago. Chang told me progress requires more than just air monitors and better technology. Just like California did, China must build a system that can track where pollution is coming from, set limits, issue permits, enforce rules—an institutional infrastructure for tackling the problem systematically. Despite the obvious differences in the Chinese and American approaches to governing, that is exactly what China is trying to create, Xie tells me. He's doing his best to help, building online tools local officials can use to assemble the information they'll need to improve their region's air. I can't help thinking that while he and so many others strive to give China the means to protect its people's health, America in the Trump era is dismantling a structure that does just that, ripping up reams of environmental rules and eviscerating the agency that enforces them. (Although Democratic California, Elaine Chang's home, remains as committed as ever.)

China has a long road to travel before it matches L.A.'s air-clearing success, and there are plenty of potential potholes along the way. But it is no India, ineffectual in the face of crisis. Chinese leaders, at long last, are doing something about the threat killing so many of their people, their course steady enough despite the zigzags to make clear which way they're headed. Change may be coming in fits and starts, but there's little doubt now that it's on the way.

* * *

Those changes have consequences far beyond China's borders. Because the actions it is taking to remove the poison from its air are the same ones that cut carbon emissions. In order to reduce its reliance on coal, it is rolling out history's most ambitious investment in renewable energy, plowing hundreds of billions of dollars into wind and solar power. Its leaders spotted opportunity, and set out to ensure China would dominate the manufacture of solar panels, just as it's done in toys, clothes, and electronics. That hope quickly became reality, and Chinese production has upended the economics of clean energy, sending costs tumbling—down 90 percent in a decade, and still dropping. From the deserts of the Middle East to the rooftops of Mexico, cheap Chinese panels are making solar power cost-competitive with natural gas, and even coal. That puts the prospect of serious reductions in greenhouse gas emissions well within reach. And China itself is at the forefront of change; by one estimate, it is installing three football fields' worth of solar panels every single hour.[22]

A short drive from Shanghai, Xiao Jin and I see this renewables revolution up close. In just one year, Jinko Solar produced panels with an electricity-generating capacity equivalent to about 10 typical coal-fired plants. The company has invited us to see how they're made, so after pulling on long blue lab coats, surgical caps, and shoe covers, we head onto the floor of one of its factories. It's clean and bright, a loud whoosh the only sound as red, robotic arms lift and twist. Solar cells—squares of silicon, lined with narrow strips of silver—come into the factory by the thousands, from another Jinko plant. Layer is added to layer, and workers in blue uniforms and baseball-style caps test and tweak as the parts are sealed together into ready-to-install modules, wires dangling from a small black box mounted on the back.[23]

Forklifts move cardboard boxes full of new panels, and they're stacked on top of wooden pallets and wrapped in plastic, waiting to be shipped for sale around the world. It occurs to me that although they are central to the issues I write about, I've never, until now, given a moment's thought to how these panels are made, or how they work. "I'm not sure how much you understand the solar cell," an expert in the company's research lab says. Laughing a little sheepishly, I suggest he assume minimal knowledge. The cells function like semiconductors, I learn: They're made of silicon that has been manipulated into two types, positive and negative, that are then layered together. Sunlight starts electrons flowing, and they are harnessed in an electric circuit.

Like the other elements of its pollution war, China's renewable power rollout has not been altogether smooth. In its sparsely populated northwest, vast wind and solar farms generate power that never gets used, because no one thought to install the transmission lines to bring it to the big cities where it's needed. That build first, think later strategy has left far too much energy going to waste. But such problems are solvable, and officials have begun tackling them. It's hard to know whether they care much about global warming, but it may not matter. These days, their actions offer one of the best hopes for staving it off.

Today, China knows moving away from coal and toward renewables is not only in its own interest, but essential to both its people's health and its long-term stability. This is what put President Xi Jinping in a position to promise, in a 2014 deal with President Obama that paved the way for the Paris climate agreement, that his country would put carbon emissions on a downward curve by 2030. And it is what will keep China moving forward even as America, under Donald Trump, heads the other way. Regardless of climate or carbon or international politics, Ma Jun tells me, "we have to do this."

It's hard to overstate the importance of that dynamic. For years, advocates of climate action pleaded with China, the world's largest greenhouse gas emitter, to join the fight. Back home, I would often hear those who cared about climate say a decision to bike instead of drive, to buy low-energy lightbulbs, felt futile next to China's relentless coal burning. Those who preferred to ignore the problem found a convenient excuse in Chinese intransigence. George W. Bush cast China's exemption from the Kyoto Protocol as reason for the United States to stay out. Even European nations willing to take more aggressive action wondered if China's vast hunger for coal would render their efforts meaningless.

In the space of a decade, that was turned on its head. China has gone from being a pretext for the rest of us to do nothing to the de facto leader of the fight against climate change. It's on its way to meeting its 2030 promise, quite possibly ahead of schedule. And other nations, particularly poorer ones hoping to learn from China's search for a cleaner, healthier way to grow, may look to its example.

Those who care about the planet's future are right to despair at Trump's rejection of science, his administration's refusal to acknowledge reality. Such intransigence makes the path to a stable climate steeper, if not impossible, to climb. But the United States is not the world's only actor. China is moving on without us.

11

"TO WHOM BELONGS THE CITY?"

Berlin Looks Beyond Cars

A gulp of air, filling the lungs. What comes next feels preordained, the relentless ticking of life's clockwork. This time, though, something steps in, jamming the gears, interrupting the rhythm. The conscious mind knows it won't win out over the body's primal force. But while it can't stop the exhalation, it can delay it—for a minute, maybe two. It won't be easy. This takes work, work that gets harder every second. As the effort of resisting grows more intense, the mind's concentration does too. And the body does what it can to protect itself. The heart rate falls. In an internal triage, vessels constrict, and blood is drawn away from the limbs, rushing instead to the brain. In the membranes of the alveoli, the movement of gases slows, easing the accumulation of carbon dioxide that nonetheless soon makes the lungs feel ready to burst. As it builds up elsewhere, discomfort turns to pain. In the abdomen, the chest. Until, at last, there is no more fighting. The mind lets go, and the body wins its inevitable victory. The breath is released, and the rhythm resumes. The urge to live, the need for air, overpowering everything.

Katharina Uppenbrink is the neighbor of a cousin of a German friend of mine. She's dressed all in black, with short, straight hair and bright red lipstick, and when I join her at a sidewalk café in her artsy neighborhood near the center of Berlin, she's chatting with the owner's inquisitive toddlers. They're poking around our table, interested in my laptop, but I just smile mutely, so lacking in German I'm unable to converse with preschoolers. Fortunately, Uppenbrink's English is perfect. She's agreed to meet me this sunny summer evening to talk about the city she's lived in most of her life—and, more specifically, how she makes her way around it.

She and her husband sold their car a few years ago, when they moved a bit closer to the city center and their son no longer needed to be driven to school. These days, she gets most places by bike. She belongs to several car-sharing services, so if she has packages to carry, or if her son, now a teenager, pleads desperately enough for a ride, she can pull out her phone and be behind the wheel, sometimes of an electric vehicle, almost immediately. Cars can be rented here for as little as a few minutes, then parked nearly anywhere to await the next smartphone-wielding customer. Uppenbrink's also planning to sign up for electric moped sharing, so she can take a spin on one of the sleek Vespa-style machines that have begun to appear around town. They look like fun, she says. "Why not use one of these instead of a car?" And of course there's always Berlin's superb public transit system—the electric trams rattling through the streets, the U-Bahn beneath them, the longer-distance S-Bahn trains, and the buses. It takes her son about 15 minutes to get to school on the train. It's a bit far, but "they're fast, the trains."

Uppenbrink and her family are not unusual. Car ownership rates are low here compared with most other German cities. That is, in part, a legacy of Berlin's divided past, when easterners couldn't afford cars, and westerners—their home a tiny island surrounded by East Germany—didn't tend to buy them because they had nowhere to go. It's a result, too, of a present in which the city, with an economy focused around creative industries, media, and government, remains considerably less wealthy than places like Frankfurt or Munich, and is full of expats and young people. (Seizing that bohemian identity proudly, a past mayor once proclaimed Berlin "poor but sexy.")

But it's not just demographic or economic happenstance that accounts for the relative dearth of cars here. Berlin is trying to do things differently, to orient itself around the needs of human beings rather than of the huge hunks of metal that hog the roads and dominate the landscape of most of the world's cities. It's luring people from behind the wheel by making other forms of transportation comfortable and convenient, while also providing a push with bans that keep the oldest, most polluting vehicles out.

Some of the Berliners I meet complain, like the denizens of so many other cities, about traffic, and it remains a problem, to be sure. But it's clear from the moment the airport train drops me in Berlin's center that some important differences mark this place out from its peers. There are far fewer cars on streets than in just about any other major city I've been to. Munching pizza at a sidewalk table in Mitte—it means "center" and, true to its name, is the capital's geographic heart— I'm in Berlin's equivalent of Midtown Manhattan or Piccadilly Circus. But the traffic trickling past is more akin to what I'd expect in a small town. Every few minutes, a car comes by, maybe a few in a row before the next lull. It's the same nearly everywhere I go in Berlin.

In young, funky Kreuzberg, the sidewalks are crowded with people late into the night—they're eating and drinking at outdoor restaurants and bars, strolling with their families, relaxing on benches beside a playground. Far outnumbered by bicycles, cars trundle past at about the same rate as in Mitte. The air is fresh and pleasant as I scarf down scallops and dumplings at a crowded table outside a popular Vietnamese restaurant, then browse the kitschy knickknacks in shop windows before planting myself on a bench in front of an ice cream shop with a double-scoop cone.

Air quality is not the only benefit I can feel. Cars take up space, create noise, make it difficult, even dangerous, for pedestrians and cyclists to move around. And of course the traffic they create saps time, productivity, and pleasure in cities and suburbs around the world. Their relative scarcity here—and it is only relative—brings the presence of something different. Not tranquility, exactly, for Berlin is vibrant and brimming with energy. But the space for a different kind of life to emerge, life on a human scale, not drowned out by the rumble of engines or choked by their fumes.

Another quality-of-life boon in Berlin is the ease of getting around. When I first moved to England, a friend told me no matter where you wanted to go in London, it would take about 45 minutes. With surprising consistency, I have found that to be mostly true. What's more, by the time I arrive, I often feel wrung out by the grime and noise of the Tube or the bus, the sheer schlep of getting anywhere. In Berlin, public transport is quick and comfortable, even at rush hour. The rule of thumb seems to be more like 20 minutes.

And that's before I hop on a bike, and the city really opens up to me. Now I can return to my rental apartment between interviews, pedal past locals drinking beer on a canal's grassy bank, and zip over to a market whose weekly street-food evening catches my eye. Cycling, of course, is gaining popularity in many developed-world cities. But it's part of ordinary life here in a way that feels different from London, New York, and many of the other places embracing two-wheeled transportation. Bikes are everywhere in Berlin—propped against benches while hungry locals pick up dinner, parked outside office buildings, and toted by parents riding trains with helmet-wearing toddlers. I rent mine from a convenience store around the corner from where I'm staying.

A biking advocate I meet uses a term I haven't heard before to refer to the group whose exclusive preserve he doesn't want cycling to be: "mammals," or Middle-Aged Men in Lycra. I value my home life, so I wouldn't want to suggest I'm married to such a creature (and anyway, Dan tends more toward old T-shirts and baggy shorts). But it's true that even my minimalist husband has acquired a daunting collection of cycling gear, one that can feel like a barrier to entry: pricey, lightweight bike, serious (expensive! heavy!) locks, fluorescent vest, and more.

In Berlin, most cyclists don't even wear helmets, let alone Lycra. They ride to work in business suits or skirts, stylish jackets, and nice shoes. Many of the bikes I see are cheap, nothing-special models. Riders cruise up to a railing or lamppost, hop off, and click the lock on in a matter of seconds, a contrast to the extensive fussing I associate with the process at home. What's more, many Berliners live in apartment buildings with big courtyards perfect for storing bikes. They simply roll through the front door and park anywhere. Quite different than my experience in London, where I keep my bike on our back deck, under a rain cover that always seems to catch on the pedals. To take it out, I have to lug it through the kitchen and down a steep flight of stairs, often bashing a shin or scuffing the wall. Just contemplating this process is tiring, so I generally end up deciding to walk instead.

Of course, the more important difference is that I'm afraid to ride on London's busy roads, where cars and trucks barrel past just inches away. Berlin's streets are wider, with far less traffic. But, trying to be fair, I remind myself how much harder it would be to make room for bikes on London's ancient, winding streets. I'm envisioning a particular two-lane road near my home, and it's a moment before I realize how much my imagination's been constrained by the cars-first design that is the norm nearly everywhere. In fact, the road I'm thinking of isn't two lanes, but four—two for moving cars and two for parked ones. We're so used to giving over our public space to automobiles that we sometimes don't even notice we've done it.

Katharina Uppenbrink agrees that here, cycling doesn't feel like a special activity reserved for the hardy few. She runs an organization that advocates for writers and artists, and she bikes to meetings at the Bundestag, Germany's parliament. "It takes me 10 to 15 minutes to get to the governmental area," for such events, "12 minutes to cycle to my office. If it rains really hard, it's three stops on the underground, about three and a half minutes," with a wait rarely more than that. Not long ago, she was invited to a dinner hosted by a high-ranking politician: "It was 12 women and I think 8 of us were cycling." No one saw any conflict between dressing smartly and riding a bike.

Uppenbrink has lived in Berlin since she was 10, with the exception of a few stints away. When she was growing up here, the wall imprisoned half the city's population, but it couldn't hold back the pollution produced by East Germany's heavy industry. So westerners like her breathed the foul smoke, too; and indeed, many of them also burned coal to heat their homes. "You could smell it and you could see it, the pollution," she says.

She was in London when the old regime crumbled, so she missed her hometown's most joyous moment, when young people danced atop the wall. But she's seen the dramatic change since then. And she's glad she's raising her son in a place so much healthier than it once was, so much healthier than other cities. She thinks part of the difference comes from the open-mindedness she feels here. "Berlin is slightly cooler, accepts different ways of life. Compared to Munich, for example: my sister-in-law, she couldn't live without a car."

I ask whether she misses owning a car, and her answer is clear. "I hated it. It's stress." Parking was always a hassle, the search for a spot sometimes taking longer than the drive. "It's so much easier with car sharing." Then, after we say goodbye, she gets up, hops on the powder-blue bike that's been sitting a few feet away from us, and—high-heeled red leather clogs notwithstanding—zooms off.

*　*　*

Berlin is no pure, pristine utopia. Not long after I arrive, the spokesperson for a national car-sharing group I've emailed wonders in his reply why I've picked this place as my clean air model.[1] For a moment, I worry I've chosen the wrong example, flown here for a chapter based on a misapprehension. But with a few sips of strong coffee, I soon rebound, and remind myself why I've come. This city's achievements, while incomplete, feel more relevant, more reachable than those I might encounter in some remote, pollution-free corner of Canada or Finland. The capital of the nation that is Europe's economic engine, Berlin is a peer to places like Paris, Rome, and London, a city whose progress holds useful lessons for the rest of us.

Yes, it has a long way to go. Despite my giddy first impressions, there are indeed traffic jams here, and a worryingly high number of cyclists are killed in accidents. Air pollution remains a problem. But in answer to the question of how good Berlin's air is, how well it has done in reducing car use, the bottom line, I've come to realize, is this: It depends on what you compare it to.

That's how Axel Friedrich sees it too. Friedrich has worked on air quality for decades, and I've planned my visit to Berlin around his availability, a tougher challenge than I'd anticipated. Since leaving his post at Germany's Federal Environment Agency in 2008, he's traveled the world as an adviser and consultant, sharing stories of Berlin's progress with those struggling to emulate it. I even saw his name on a report someone handed me in China. His emails, as I attempt to pin down my trip dates, announce his latest stop: "Greetings from Malta," he writes, "from Washington, DC," "from Lisbon." When we finally meet, he says he's leaving the next day for the Greek island of Corfu for some work on pollution from shipping.

The trek to Friedrich's house breaks my 20-minute Berlin rule of thumb. He lives on the outskirts of town, and it's more like an hour by train and bus before I'm walking along suburban-feeling streets lined with big trees, the grassy verges on either edge of the sidewalk overgrown with wildflowers and weeds. His apartment on the top floor of a big house is all white, lined with tall windows, glass doors, and skylights, but the sparse, modern effect is undercut by the mess piled on every available surface: papers and magazines, tubes of glue, old laptops and phones and chargers. Friedrich is in wrinkled tan trousers, a baggy sweater, and furry black slippers. His white hair is brushed forward, his broad face craggy and angular.

He's not blind to Berlin's shortcomings. Air here is among the best of Europe's big cities, he boasts, "but I am not happy. Still fighting on a number of things." The latest battle is with the national government, over a sticker system Berlin wants to introduce to help it recognize diesel cars that are truly clean, so it can bar the rest from entering. But while he's always pushing for more, he's proud of his city's successes. "If I'm here, it's bad," he laughs. "If I go to other cities, I say, 'Berlin is an example.'" Compared with London, he says, this place "is a heaven." He believes his city's achievement can show others it's possible to combine clean air and high quality of life with a thriving economy. "You need always examples," he says. "Berlin for me is a showcase."

Political leaders, he says, are accustomed to making arguments with "hard facts, economic facts. But many things are soft items"—he lifts a hand and gently squeezes his fingers together—"which people don't really see, but they feel." The livability that comes from Berlin's green spaces, its progress on pollution and noise, he believes, can lure the talent that helps a city thrive: "If you have a family, do you go to a dirty city?"

Friedrich tells me there are concrete steps cities can take to clean up. He credits much of Berlin's success to a plan devised at the start of the century. Its biggest strength, he says, lay in combining three sorts of goals—economic, social, and environmental—to create a long-term vision for the future. The question at its heart is one facing urban areas around the world: "To whom belongs the city?" is how he sums it up. "Cars have taken over our cities. And we are trying now to recover the space for humans."

Cities need trucks and vans, of course, to bring in goods and remove waste. Taxis add a useful link to transport networks. But it makes no sense, in his view, for individuals in private cars to get so much of a commodity—space—that's in such high demand. In addition to the pollution and noise they create, those cars also chop up neighborhoods, endanger pedestrians, and create obstacles to walking. "All the things which we as humans need are destroyed."

It wasn't just one decision that helped get Berlin where it is today, of course. There have been many choices along the way. And if public transportation is good enough to pick up the slack, Axel Friedrich thinks one of the best ways to get people out of cars is simply to restrict parking. "If you go today in the center of Berlin, you have to be an optimist if you go by car," he says. Encouraging housing and businesses to develop near one another helps reduce commuting time. But the most important lesson Berlin can offer, in his view, is the importance of big-picture thinking, of setting out a vision, making a plan, and then carrying it out. In 2000, Berlin was thinking hard about what kind of place it wanted to be decades hence. Without that vision, smaller changes just amount to tinkering around the edges. "If you have no goals, where do you walk to?" he asks. "You need directions."

* * *

Martin Lutz has been part of the change here too. A trim man, energetic and warm, with close-cropped gray hair and beard, he's spent years as an air expert for the city government. It's early morning when I poke my head into his office in a nondescript modern building and take in a view that encompasses Berlin's spiky, iconic TV tower, the elegant dome and spires of a nearby church, and the Spree River, bounded by stone walls. When Lutz first came here as a student, East German power plants and heavy industry were burning vast amounts of lignite, a particularly dirty grade of coal, mined from shallow pits. Those dinosaur-like factories began to close after reunification, and the old power stations were modernized or replaced, so pollution levels fell fast.

More surprising, Lutz says, has been the city's ability to hold on to a more positive legacy of its painful past. As a westerner in the Cold War years, "you didn't need to have a car," because no one could leave the city without a visa. So when the wall fell, only about 300 of every 1,000 inhabitants owned one. "I would have bet all sorts of bottles of champagne that this would change" in the years that followed, Lutz tells me. "And I would have lost," because in nearly three decades, it's climbed to only 342 cars per 1,000 Berliners.[2] "This is surprisingly fantastic," he says. In Frankfurt, by comparison, 541 of every 1,000 inhabitants own a car;[3] in Cologne, it's 425.[4] Berlin's achievement comes in large part from the investment poured into knitting its divided public transit network back together. After the wall came down, abandoned stations were reopened, lines that had stopped at the border were reconnected, and the entire system was expanded and upgraded.

Lutz has been central to other changes that kept the air improving too. Berlin began retrofitting all its diesel buses with particle filters starting in 1999. The city doesn't have the power to require private construction projects to use equipment with the cleanest engines, but it demands them when it hires for public work. It upgraded police and fire vehicles, and garbage trucks meet strict standards, too, many of them running on biogas made from kitchen waste. And in 2008, the city introduced its Low Emissions Zone, barring the oldest and most polluting vehicles from the center.

Pollution, though, can drift a long way. Lutz mentions Polish coal, a bane to so much of Europe, and German power plants still burn the stuff too. So there are limits to what Berlin can do on its own. And Martin Lutz is furious about another obstacle that has thwarted progress. The Low Emissions Zone was built on what turned out to be a shaky foundation: the pollution limits car companies were flouting. Their cheating made a mockery of Berlin's efforts, Lutz tells me. With cars so many times dirtier than they purported to be, the restrictions became meaningless. He holds his thumb and forefinger close together to show how little the zone did to lower nitrogen dioxide levels. Officials wanted to encourage residents to move to cleaner, more modern cars, but it turned out the ones drivers were being nudged toward were hardly better than the ones they were ditching. "If the new vehicle stock coming into the fleet is not better than what you try to get rid of," Lutz says with a bitter laugh, "there is no improvement."

Even now, he complains, Germany's government seems more concerned about the car owners hoodwinked into buying rule-breaking models than the people who, in far greater numbers, must breathe the consequences. He wants all the cheating cars retrofitted with treatment systems that bring them into line with the law, not just the less expensive software tweaks industry would rather provide. And it's clear to him who should pay: the companies responsible. But he's under no illusions about the likelihood of that happening. "Nice joke, isn't it?"

Dieselgate, it seems, is the story I can't get away from here. While I ask those I meet about bike lanes and clean air and quality of life, their answers always seem to circle back to what many see as a brutal betrayal by the car companies that are pillars of their country's economy. Indeed, Lutz had to reschedule our meeting around a Bundestag hearing about the cheating's fallout.

Today, the anger the scandal has set off looks as though it may be coming back to bite those responsible, or at least to wrench them away from the dirty fuel they pushed for so long. Mayors don't have the power presidents and prime ministers wield, but across Europe, many are drawing up ever-stricter plans to bar diesels from their streets. And it turns out that limited step may reverberate well beyond cities' limits. Because who wants to buy a car that may not be allowed to take you where you want to go? So, for the first time since diesel sales began booming just after the turn of the century, they are now on the decline. Although it will be a long time before all those old cars are off the roads, the fervor of Europe's embrace of this dirty fuel may be cooling once and for all.

* * *

Like Martin Lutz, Peter Feldkamp wants Berlin to be better. He believes the city hasn't always done enough to turn its admirable visions into reality. Feldkamp grew up going on cycling vacations with his parents and younger brother, but he didn't think about biking in political terms until much later. Now, he says, he understands that decisions made by politicians help determine how safe he feels on the road. He's young and slim, with dirty blond hair and a stubbly goatee, the sleeves of his khaki shirt pushed up. Like many I meet in this city, he tells me Berliners' enthusiasm for bicycling has flowered despite a dearth of protected lanes and other infrastructure to make it safe and easy. Even here, he says, officials are far too focused on the needs of drivers. "They're still building a car-centric city."

In Germany, he complains, making room for bikes often means little more than painting white lines on the road. There are few of the concrete barriers that have begun springing up in other cities to separate cycling lanes from traffic. Feldkamp tells me he envies the protective curbs that have turned Manhattan into a place where even young children can ride their bikes. In Berlin, he regularly attends memorials for cyclists killed in accidents; their fellow riders block the road where it happened and sit on the ground in silent tribute. Just a few weeks ago, he says, he went to one for a man named Michael, killed in his mid-forties when a driver opened a car door in his path. "It really is about life and death. If there would have been good infrastructure, Michael would still be alive."

Plenty of others, Peter Feldkamp tells me, share his frustration at their city's failure to do more to encourage a form of transportation with the potential to bring big reductions in both traffic and pollution. A few years ago, he and a group of fellow activists began pressing for a referendum on a raft of measures to make Berlin safer for cycling. As the first step toward getting it on the ballot, they had to gather 20,000 signatures in six months. "It took us three and a half weeks. We collected 107,000." Standing on corners asking passersby for support, he laughs, was like giving away free beer. Once, he was taking a break, talking to a friend, when a man came up and asked if he was with the bike campaign. "I've got signatures for you," the guy said, pulling out a stapled bundle of papers covered with scrawls he'd assembled. When the campaign put up a web page with an email address, 150 messages arrived within hours. "Nearly every single email contains this one word: 'Finally,' " Feldkamp recalls. "Politics just ignored this movement for so long, and people got so angry."

For all the cyclists I've seen on Berlin's roads, Feldkamp thinks there are many more who want to join them but don't feel comfortable doing so. Even he is unnerved when an impatient bus driver starts creeping up behind him. "Imagine my 60-year-old mother cycling there. She would be very scared, for a reason. And that's the group we have in mind," he says. "My mother, maybe your kids, people who are not so experienced."

The campaign's demand was for wide, safe lanes not just on major thoroughfares within the city, but also on the roads that lead from the outskirts to the center. That, Feldkamp says, is how you make it possible for people who now drive to cycle instead. To hammer home their point, "to show we need to protect," the campaigners used a photo of a girl riding in a painted bike lane, stuffed animals lined up along its edge. In another image, kids on bikes, trailing yellow balloons, pedaled along a street empty of cars. Viewers, he said, were thrown off balance by a scene that should feel normal but doesn't. "Parents do not send their kids out to play in the street. Everyone understands why," he says. "We lock our kids in our homes so the cars can play outside."

The campaigners also wanted more parking for bikes, the redesign of dozens of dangerous intersections, and police officers patrolling on two wheels, so they're attuned to riders' needs. In the end, the referendum didn't happen. Municipal elections brought a more sympathetic coalition to power, and the activists decided to work with them on a new law rather than pursuing the public vote. Peter Feldkamp believes something better is really on the way now. And he's started advising groups making similar demands in other German cities. "It's becoming a national movement," he tells me. "It works in Berlin, and people say, 'Why shouldn't it work here?' "

* * *

Jörg Welke likes to think of Berlin as a laboratory, too, a testing ground for ideas with the potential to change the way all of us get around. We've arranged to meet for lunch at a Persian-Moroccan-Israeli place he's suggested, on the top floor of a hip hotel beside the city zoo. My phone tells me it's 17 minutes by bike from where I'm staying, but it turns out Google's estimates don't account for how often I have to stop to check the route, or for my ability, even with GPS, to get lost. So I'm late, and a little sweaty, as I hurry into the buzzing restaurant, whose floor-to-ceiling windows offer a stunning view over the city. Down below, creatures that look from this height like monkeys and deer lounge in their enclosures in the zoo, and a bit farther out, Tiergarten park stretches in a long swath of green toward Brandenburg Gate and the glass dome of the Reichstag.

Welke is tall, in dark jeans and a peach shirt, his head shaved. While I peruse the menu, he explains that the government-backed agency for which he is a spokesman works differently than I'd imagined. Rather than simply promoting electric vehicle use in Berlin, it tries to spur the growth of local companies whose ideas may help to revolution-ize transportation all around the world. And although it's called the Agency for Electromobility, its vision is broader than just electrifi-cation. Berlin's leaders believe technology is ready to revolutionize transportation, and they want their city to be a hub for the industry that makes it happen.

In the years to come, Welke believes, mobility will increasingly be sold as a service, not a product. This is an idea I've heard before, prof-fered by many who hope to loosen cities' gridlock, clean their air, and ease pressure on the climate. We'll be less likely, the thinking goes, to buy our own cars than to pay for whatever form of transportation we need when we need it—a shared bike at the end of a train ride, a self-driving car home from the supermarket. And indeed, the millen-nial generation, at least in wealthier nations, seems less interested in driving than its elders. In both America and Britain, the proportion of young people holding driver's licenses has declined steadily since the 1990s.[5] "If you own a car, it stands [idle] 23 hours a day," Welke says. "With the help of smartphones, with the help of I.T. and digita-lization and automation, we don't need that anymore."

Welke is no stranger to ideas that seem a little utopian: he once lived for more than a year in a demonstration house meant to show off the latest in super-efficient technology. As I dig into a tasty vegetable, egg, and tahini dish, he enthuses about his favorite way of getting around Berlin: on his electric bike. Although it's expensive, he says, the e-bike has become a big seller "because it convinces through its performance. It's just a perfect product." The motor makes it feasible to ride to places too far, or too hilly, to reach under his own power—and to arrive unrumpled.

For him, it's just one piece of a mix-and-match approach to getting around, a style he believes will grow increasingly common. "I just take the form of mobility that I need." Often, that's his e-bike, but this morning he came into work on a shared electric scooter, one of the Vespa-like models Katharina Uppenbrink is getting ready to try. They're free-floating rentals, meaning they have no set parking spots; riders can pick them up and drop them off anywhere, paying with their phones. This afternoon, Welke will take a bus to pick his son up from a guitar lesson, then go by scooter to meet the rest of his family at the theater. After the show, they may take what he calls a "clever shuttle," a new electric taxi service that pools passengers' journeys with those of others going the same way.

Welke rattles off some of the projects under way in companies his agency has supported. One firm shifted from making electric motors for car windows to motors for electric bikes, and now exports them as far as China. Another developed an app—he shows me on his phone—that combines data from various companies to show all the cars and scooters available right now to rent nearby, saving a user the trouble of checking each one individually. Another app displays car-charging spots around the city, with a green pointer indicating availability.

Instead of running buses on predetermined routes with set schedules, he goes on, technology can enable customized shuttle services that run when and where riders need them. Traditional bus routes work well in big cities because there are generally plenty of passengers. But for less densely populated places, a smarter service might bring new options. "Transport on demand," he calls it, and one of his companies designed an algorithm to make it happen. Berlin, with its wide roads and open spaces, its forward-looking population accustomed to life without cars, is a perfect place for experimentation, Welke says. Personally, he sees climate change as a more pressing reason than air pollution for a new approach to transportation. But the immediate health threat posed by dirty air seems to get more traction with the public, and that's fine by him.

He's invited a friend to come meet me after lunch, so we leave the restaurant for a chic watering hole right across the corridor. Theresa Theune couldn't look more out of place in a bar whose tone I got a hint of from the sign downstairs listing the DJ lineup and the neon lighting in the elevator. Warm and effusive, she wears her gray hair pulled back in a ponytail and is holding a rope attached to the shaggy, spotted dog who sits at her feet. We head out to the terrace to escape the noise inside, and she finds us a quiet spot at the end, overlooking a bombed-out church left in ruins as a World War II memorial.

As her dog settles in for a nap, Theune tells me she left Berlin in 1990, just as it was reopening to the world, because its awful air was making her children sick. They moved to a little village near Bonn, but missed home so much, they came back in 2000. Now she lives on the city's outskirts, and she didn't feel right about the gas car she was using to go to the theater or visit her grandkids. Not just the pollution it was creating, but the noise too. "It's a normal car, but every car is noisy," she says. "Nobody thinks about this." Gradually, that nagging feeling crystalized into a realization that "I'm part of the problem," so she began hunting for a used, affordable electric car.

She'd really rather cycle. But while her aging pet can still run along-side as she pedals, he wouldn't be able to keep up with an e-bike, and she's afraid to leave him home alone. The little plug-in car she bought "is more of a moving doghouse," she jokes, and she loves zipping around in it. And it was fun, one recent weekend, to watch her well-off siblings and their spouses clamor for turns behind the wheel of her car. "Everybody who drove it has a smile on his face," she says. In that way, it's a bit like an electric bike, which, she agrees with her friend Jörg Welke, elicits joy in its riders. "This is the e-bike smile, it's such a wonderful feeling." She plans to get one after her dog dies.

The truth, she believes, is that it doesn't really make sense anymore for so many of us to own cars. "Most of the time, they just transport one person. And you have one person and one and a half tons of metal and plastic." Even if fossil fuels were to be replaced by clean electricity, "you have the same accident problem, you have the same space problem." With self-driving vehicles on the way, to be ordered up, perhaps someday soon, with the tap of a screen, she can imagine a radically different future, of cleaner vehicles, fewer of them, not owned but shared.

It may not be a pipe dream to think transportation is ripe for the kind of technological disruption that has transformed so many other areas of modern life. Not so long ago, video calling seemed like a futuristic fantasy, but now we Skype and FaceTime while walking down the street. So why are we still driving cars that are little more than updated versions of the ones our grandparents had? There's so much at stake—not just our health and that of the world we inhabit, but the very ability of our cities and suburbs to function, of people to get where they need to go instead of wasting hours in maddening traffic jams.

Of course, Theune worries big car companies, and the politicians who protect them, may stand in the way of change. Like so many I've met here, she's outraged by the companies' gall in evading the rules for so many years, and by her government's failure to stop them. She struggles for words strong enough to convey these feelings. "Not only angry. Angry is too weak," she says. "Not disturbed. Furious. And let's say shocked."

She feels sure, though, that we're on the verge of something new, a fundamental transformation. Soon, she muses, it'll be hard to believe we once got around in "loud and stinking cars with a little fire inside. It will be like dinosaurs." Theresa Theune is getting older now, so she doesn't know if she'll live to see the new world she imagines. But she thinks I will. "Things are going to change," she declares. "I'm really convinced." I don't know if I share her faith. I do believe a different, better future is possible. It's up to us whether we find our way there.

EXHALE

What Comes Next

Crisscrossing the world on an itinerary built around pollution, I've listened, in a half-dozen time zones, to stories of dirty air's effects spilling forth in nearly as many languages: Mandarin and Hindi, Polish and Spanish and English. I've seen the depredations of terrible pollution, the confluence of political and economic power that too often allows it to persist, and the despair it can bring—the despondent shrugs of people who no longer believe those in charge care what happens to them. Stewing in London's diesel mess, I've sometimes felt that hopelessness myself.

But I've also found cause for optimism. Leading utterly different lives on opposite sides of the globe, people fighting the same fight have invited me in to watch it unfold. Doing what they can to ensure their families and their neighbors and perfect strangers can fulfill that most basic human need—the need to breathe, and to breathe air that isn't poisoned.

Of course, individuals can't solve this problem alone. Only our governments have the power to stop polluters who've shown, again and again through the years, that they'll put profit over human lives unless we force them to change.

In a sunny office in upstate New York, Tom Jorling handed me a copy of a law that did just that, the one he helped draft with his old friend Leon Billings in the months just before I was born. A law that, for all its technical language, is, at bottom, a statement of values, a declaration of what we are prepared to accept and what we will not. The Clean Air Act of 1970 was voted into the statute books by a generation of public servants all but passed now, a generation that included, among others, a Republican senator named Howard Baker, so proud of the part he played in its passage that he later said he'd be honored to have it etched on his tombstone.[1]

That generation's gift to us, their children and grandchildren, is air far cleaner than it once was, far cleaner, because of their work, than the stuff I breathe in London. It's a living inheritance, though, a legacy that must be defended and extended if we are to bequeath it to our own children. Instead, many Americans have come to take it for granted, forgetting the long, hard battles that brought us where we are. And the ever-present truth that what's been won can also be lost, that the gains of the past offer no guarantees for the future. While clean air feels like a birthright, it can disappear in a puff of smoke if the rules created to protect it, and the enforcement that gives those rules teeth, are unraveled.

As we absorb that painful reality, a rising power on the other side of the globe is searching for a way to replicate the achievements of an American agency that was, not so long ago, the envy of the world, but has more recently suffered an onslaught of presidential hostility, emerging weakened and undermined: the EPA. Through the choking filth they still endure, China's leaders and its people can see what we, with our clearer skies, now struggle to keep in focus: the value of a public guardian, guided by science and empowered by the law, to protect human health and draw lines polluters may not cross. India lags further behind, but perhaps someday its people, and their neighbors across South Asia, and others far beyond, will have such a protector, too, and be able to breathe air that's not horrendous.

There's another lesson in our history, one that offers a valuable signpost for today: The benefits of cleaner air almost always dwarf its costs. When we're contemplating change, the price tag tends to loom larger, and those who will have to pay often exaggerate it, hoping we'll shy away from action. But when polluters are forced to clean up, they buckle down and find the cheapest way to do it. That determination often brings innovations that make change quicker and easier than predicted. And while the cost is less than we'd feared, the benefits are often much larger, multiplying in a cascade of well-being and rising productivity. Spread among millions of people, they can be hard to see, but that doesn't make them any less real.

After all my travels, I can see now what I couldn't when I started. In the suffering pollution brings, there is also the glimmer of a different future, its outlines visible through the haze. Because as we come to grasp dirty air's dangers, there is something else we should understand: This is not an insoluble puzzle, a problem to which we must resign ourselves. We know how to fix it. If we do, we'll reap the rewards, in lives saved and health improved, almost immediately. That's what scientists like Ed Avol and Jim Gauderman have shown us, with help from the thousands of children whose breath they measured, year after year, in school gyms and lunchrooms. With every new study, they and others like them remind us why we must move forward, not back.

Yes, there are small-scale fixes, simple and necessary. Like the filters ready to snap onto smokestacks of ships moored in a Californian harbor, or the dockside electricity that lets them still idling engines. Like New York City's push to rid itself of the dirtiest heating oil when it saw that outdated furnaces in just 1 percent of buildings were creating more soot than all its traffic. And bigger solutions, too, like the rejection Britain, and all of Europe, must finally deal to diesel.

Such changes will bring meaningful improvement: fewer heart attacks, less cancer and asthma and dementia. But the real answer, the one with the power to bring truly healthy air—along with another, even greater prize—lies in a more fundamental change. A shift, at long last, away from fuels that, while they poison our bodies, are also wrecking our planet.

Looming over the air pollution crisis, of course, the frightening backdrop to the stories of this book, is the existential threat of climate change. We're already feeling its effects, in weird, unseasonable weather, vicious storms, and temperatures that seem to be always shattering records. The warning lights are flashing, more urgently than ever. But for most of us, this is a danger that still feels abstract, distant. Not as pressing as one that hits us where we live, threatening our very bodies and those of our children and parents.

That, it turns out, is where the good news multiplies. Because the changes that will lengthen, and improve, lives in the here and now are the same ones with the potential to save us from the most catastrophic dangers of runaway warming. For these two crises are deeply intertwined, both symptoms of the unhealthy foundation on which we have built our world: fossil fuels. Both will abate only when we replace those fuels with something better. When we finally end the reign of coal, which has brought industrialization and prosperity in so many places, but whose dark side has blighted lives for far too long now, and on which we no longer need to rely. And wean ourselves as well from gasoline and diesel, and the oil from which they come. A revolution, too, in the way we get around: cleaner cars, and fewer. The building blocks of a different kind of world, healthier and more resilient.

The change we need now is more dramatic, more radical, than the steps we've taken in the past. The price of not making the leap is greater than before, but the rewards will be too. Ending millions of lives every year and blighting many more, air pollution is cause enough for action. But the imperative of confronting climate change means something even bigger is at stake: our very survival, on a planet capable of supporting us, one where floods don't drown our biggest cities and drought doesn't parch the fields that feed us.

While it can be difficult to imagine a world different from the one we know, it is within our reach. The unconscious mind controls breathing and so many of the body's other vital functions, but the decisions that determine the parameters of all our lives ought to be made consciously. We have a choice: We don't have to give our cities over to the cars crowding their roads, don't have to let the ships that underpin our global economy poison the bodies of those who live along coasts, don't have to accept that simply stepping outside can sap our strength.

I believe we have it in us to rise to the challenge, to create something better than what we were given. In the end, it's up to us. We hold the power to build a cleaner, healthier future. One in which breathing, life's most basic function, no longer carries a hidden danger.

* * *

Deep inside the body, the oxygen delivered by the breath is on its way to the muscles, the organs, the brain. The carbon dioxide they must be rid of is filling the lungs, ready to make its escape. As the muscles of the chest wall relax, the ribs fall inward, and down, and the diaphragm springs into a dome. The chest is too small to hold all that stale air now, and the tightening confines push it back out. The breath retraces the path it took seconds earlier, whooshing upward through the bronchi, over the voice box, then out. Borrowed only for a moment, now returned. Released, again, into the world.

There's just an instant's pause before the cycle begins once more. No rest, as the body demands the thing it cannot do without. In and out, again and again, hour after hour, day after day. From the moment of birth until our time reaches its end.

Life. Breath. Air.

ACKNOWLEDGMENTS

More than 250 people shared their time, their experiences, and their expertise with me over the three-plus years I spent reporting on air pollution, in many cases long before it was clear this project would actually become a book. Some are quoted in the preceding pages; others are not, but my conversations with them nonetheless helped shape my understanding of this issue, so their contributions are, in a very real way, part of *Choked* too. I am grateful to all of them, and I've done my utmost to convey their thoughts accurately and do justice to their stories. This book wouldn't exist without them.

I owe thanks to the Society of Environmental Journalists' Fund for Environmental Journalism, which supported my trip to India. I'm also deeply grateful to the Pulitzer Center on Crisis Reporting, which funded my research in both Poland and China and has championed the work that came out of those travels. Much of my research would have been impossible without the backing of the SEJ and the Pulitzer Center, but the value of the grants they gave me was not only monetary. Their endorsement of my belief that the story of air pollution was an important one that needed telling gave me the confidence to keep pushing forward. At a time when journalism's business model is fracturing, but in-depth, independent reporting is more necessary than ever, both these important organizations have stepped into the breach. I'm proud to have worked with them.

Reporting from my air pollution travels has appeared, in various forms, in the *Guardian, National Geographic* online, and what was then the *International New York Times*. I am grateful to the editors who supported my grant applications and made my stories better: Katherine Knorr, Brian Childs, Anne Bagamery, and Stanley Reed, then at the *INYT*; Mike Herd, Nick Van Mead, and Chris Michael at the *Guardian*; Robert Kunzig, Brian Howard, and Victoria Jaggard at *National Geographic*. My wonderful former Associated Press colleagues and friends Beth Harpaz and Jean Lee also lent support as I applied for funding.

I couldn't have navigated my international reporting without the help of fixer/translators who were also journalistic partners, excellent travel companions, and, before long, friends: In China, Xiao Jin and I have agreed to withhold her full name, but she has my heartfelt gratitude. As do Marcin Krasnowolski in Poland, and Neha Tara Mehta and Ravi Mishra in India. Neha also took beautiful photos that accompanied my published articles. My Delhi driver, Amit Kumar, jumped enthusiastically into the project, too; although I have a couple of decades on him, I gladly clasped the parental hand he offered as we darted across a busy highway (undoubtedly the most terrifying moment in my reporting). Katy Daigle provided invaluable India advice. TERI, The Energy and Resources Institute, helped arrange my trip to Mustafabad. In Los Angeles, the Sloan-Decters—Ben, Jackie, Addie, and Leo—gave me a place to stay (and a tour of the Children's Ranch!).

I'm so grateful to the University of Chicago Press for its enthusiastic embrace of this project, and particularly to Karen Merikangas Darling, whose faith in *Choked* helped make it a reality. She has been a tireless champion of the book, and I appreciate her advocacy, the counsel and support she offered, and her incisive editorial suggestions. Susannah Engstrom helped shepherd *Choked* through production, Norma Sims Roche saved me from many errors with her careful copyediting, and Erin DeWitt oversaw the editing process, further polishing the text. Sincere thanks also go to Melinda Kennedy, for working hard to make sure *Choked* would reach as many readers as possible.

It's been a joy to work with Laura Barber and her fantastic team at Granta Books in Britain. Laura provided cheerful (and much-needed!) encouragement through months of writing, and wise guidance that helped me see the changes my manuscript needed. Her input made this book stronger, and I count myself fortunate to have worked with such a gifted editor. Ka Bradley read later drafts and guided *Choked* toward its U.K. publication when Laura went on maternity leave; Christine Lo brought my words onto the page. Lamorna Elmer, Pru Rowlandson, Natalie Shaw, and Simon Heafield helped this book find its British audience, and I'm so thankful for the energy and skill with which they did so.

I couldn't have asked for a more wonderful literary agent than Jessica Papin, whose unflagging advocacy got this book out into the world and whose expert eye improved it at every turn. I'm so glad to have her as a guide through the publishing labyrinth, and as a friend too. Thanks as well to the rest of the Dystel, Goderich & Bourret team, especially Lauren Abramo, whose work on international rights means I get to see my words rendered in languages I will never speak.

I'm grateful to the Society of Environmental Journalists' mentorship program for connecting me with Lisa Palmer, who shared her experiences of the publishing world and introduced me to Jessica. The members of my on-and-off writing group—Carla Power, Ginanne Brownell Mitic, and Camilla Bustani—read my proposal and early chapter drafts, and their comments helped me climb the learning curve of book writing.

I'm so grateful for my parents, Ronnie and Barry Gardiner, who have always been a source of boundless love and encouragement. Their belief in me has underpinned everything I've done, and their enthusiasm for this book helped keep me going. I'm lucky to have Jill Noonan as my sister and friend, and to feel our bond deepening with every year.

Anna Waldram never let me forget there was life outside this book; she's grown from a young child to a preteen as it's progressed. Her curiosity, spark, and sense of fun brighten every day, and I'm so proud to be her mom. Dan Waldram has spent countless hours discussing every aspect of *Choked*, reading multiple drafts, and offering ideas that are woven through it. He put up with my grumpiness when the work wasn't going well (and when it was), held down the home front while I traveled, and always believed I had a book in me, even when I wasn't so sure. He's my best friend, true love, and essential support.

NOTES

Prologue

1   Aristotle, "On Youth and Old Age, On Life and Death, On Breathing," Internet Classics Archive by Daniel C. Stevenson, Web Atomics, accessed March 13, 2018, http://classics.mit.edu/Aristotle/youth_old.mb.txt.

2   World Health Organization, "9 out of 10 People Worldwide Breathe Polluted Air, but More Countries Are Taking Action," news release, May 2, 2018, http://www.who.int/news-room/detail/02-05-2018-9-out-of-10-people-worldwide-breathe-polluted-air-but-more-countries-are-taking-action. This toll includes the effects of household air pollution, caused mainly by dirty cooking stoves in poor nations.

3   Kimberly Chriscaden, communications officer, World Health Organization, email message to author, September 25, 2017.

4   Richard Burnett et al., "Global Estimates of Mortality Associated with Long-Term Exposure to Outdoor Fine Particulate Matter," *Proceedings of the National Academy of Sciences of the United States of America*, September 4, 2018, https://doi.org/10.1073/pnas.1803222115.

5   American Lung Association, *State of the Air 2018*, accessed August 20, 2018, http://www.lung.org/assets/documents/healthy-air/state-of-the-air/sota-2018-full.pdf.

6   Heather Walton, David Dajnak, Sean Beevers, Martin Williams, Paul Watkiss, and Alistair Hunt, *Understanding the Health Impacts of Air Pollution in London*, report prepared for Transport for London and the Greater London Authority, July 14, 2015, p. 8, http://www.kcl.ac.uk/lsm/research/divisions/aes/research/ERG/research-projects/HIAinLondonKingsReport14072015final.pdf; "Deaths Registered by Area of Usual Residence, UK," Office for National Statistics, January 25, 2017, http://www.ons.gov.uk/peoplepopulationandcommunity/birthsdeathsandmarriages/deaths/datasets/deathsregisteredbyareaofusualresidenceenglandandwales.

7   In 2014, there were 399,000 deaths in the EU-28 nations from PM2.5 exposure alone, according to European Environment Agency, *Air Quality in Europe—2017 Report*, p. 9, https://www.eea.europa.eu//publications/air-quality-in-europe-2017. Deaths from nitrogen dioxide and ozone exposure would push the total even higher, but it is unclear whether the effects are

additive. In 2014, there were 26,000 road fatalities in Europe, according to European Commission Directorate General for Mobility and Transport, "EU Road Fatalities," February 2016, accessed November 13, 2017, https://ec.europa .eu/transport/road_safety/specialist/statistics_en.

8   Janet Currie and Reed Walker, "Traffic Congestion and Infant Health: Evidence from E-ZPass," *American Economic Journal: Applied Economics* 3, no. 1 (2011): 65–90, doi:10.1257/app.3.1.65.

9   Joseph Lyou, president & CEO, Coalition for Clean Air, El Segundo, California, interview with author, January 28, 2015.

Chapter One

1   W. James Gauderman et al., "The Effect of Air Pollution on Lung Development from 10 to 18 Years of Age," *New England Journal of Medicine* 351, no. 11 (2004): 1057–67, doi:10.1056/nejmoa040610.

2   Health Effects Institute, *State of Global Air 2018*, Boston, Massachusetts, 2018, p. 3, https://www.stateofglobalair.org/sites/default/files/soga-2018-report.pdf.

3   World Health Organization, "Ambient Air Pollution: A Global Assessment of Exposure and Burden of Disease," Geneva, 2016, p. 15, http://apps.who.int/iris /bitstream/10665/250141/1/9789241511353-eng.pdf?ua=1.

4   "Explore the Data," State of Global Air 2018, accessed June 6, 2018, http:// stateofglobalair.org/data/#/health/plot.

5   Massachusetts Institute of Technology, MIT News Office, "Study: Air Pollution Causes 200,000 Early Deaths Each Year in the U.S.," news release, August 29, 2013, accessed June 6, 2018, http://news.mit.edu/2013/study-air -pollution-causes-200000-early-deaths-each-year-in-the-us-0829.

6   World Health Organization, Media Centre, "7 Million Premature Deaths Annually Linked to Air Pollution," news release, accessed November 9, 2017, http://www.who.int/mediacentre/news/releases/2014/air-pollution/en/.

7   World Health Organization, "9 out of 10 People."

8   "GBD Compare," Institute for Health Metrics and Evaluation, 2018, accessed September 11, 2018, https://vizhub.healthdata.org/gbd-compare/.

9   Burnett et al., "Global Estimates."

10   The World Bank and the Institute for Health Metrics and Evaluation of the University of Washington, Seattle, *The Cost of Air Pollution*, Washington, DC, 2016, p. vii, https://openknowledge.worldbank.org/bitstream/handle /10986/25013/108141.pdf?sequence=4&isAllowed=y.

11   "Ambient (Outdoor) Air Quality and Health: Fact Sheet," World Health Organization, updated September 2016, accessed November 9, 2017, http://www .who.int/mediacentre/factsheets/fs313/en/.

12   R. D. Brook et al., "Particulate Matter Air Pollution and Cardiovascular Disease: An Update to the Scientific Statement from the American Heart Association," *Circulation* 121, no. 21 (2010): 2331–78, https://doi.org/10.1161 /CIR.0b013e3181dbece1.

13   Anoop S. V. Shah, Kuan Ken Lee, David A. McAllister, Amanda Hunter, Harish Nair, William Whiteley, Jeremy P. Langrish, David E. Newby, and Nicho-

las L. Mills, "Short Term Exposure to Air Pollution and Stroke: Systematic Review and Meta-analysis," *BMJ* 350, no. H1295 (March 24, 2015), https://doi.org/10.1136/bmj.h1295.

14 Kristin A. Miller, David S. Siscovick, Lianne Sheppard, Kristen Shepherd, Jeffrey H. Sullivan, Garnet L. Anderson, and Joel D. Kaufman, "Long-Term Exposure to Air Pollution and Incidence of Cardiovascular Events in Women," *New England Journal of Medicine* 356, no. 5 (2007): 447–58, doi:10.1056/NEJMoa054409.

15 R. D. Brook, "Inhalation of Fine Particulate Air Pollution and Ozone Causes Acute Arterial Vasoconstriction in Healthy Adults," *Circulation* 105, no. 13 (2002): 1534–36, https://doi.org/10.1161/01.CIR.0000013838.94747.64.

16 Andrew J. Lucking et al., "Diesel Exhaust Inhalation Increases Thrombus Formation in Man," *European Heart Journal* 29, no. 24 (December 1, 2008): 3043–51, https://doi.org/10.1093/eurheartj/ehn464.

17 Joel D. Kaufman et al., "Association between Air Pollution and Coronary Artery Calcification within Six Metropolitan Areas in the USA (the Multi-Ethnic Study of Atherosclerosis and Air Pollution): A Longitudinal Cohort Study," *Lancet* 388, no. 10045 (August 13, 2016): 696–704, http://dx.doi.org/10.1016/S0140-6736(16)00378-0.

18 C. Arden Pope III, A. Bhatnagar, J. P. McCracken, W. Abplanalp, D. J. Conklin, and T. O'Toole, "Exposure to Fine Particulate Air Pollution Is Associated With Endothelial Injury and Systemic Inflammation," *Circulation Research* 119, no. 11 (2016): 1204–14, https://doi.org/10.1161/CIRCRESAHA.116.309279; Nicholas Bakalar, "Air Pollution's Toll on Heart May Begin Early," *New York Times*, October 25, 2016, https://www.nytimes.com/2016/10/25/well/live/air-pollutions-toll-on-heart-may-begin-early.html?_r=0.

19 C. R. Jung, Y. T. Lin, and B. F. Hwang, "Ozone, Particulate Matter and Newly Diagnosed Alzheimer's Disease: A Population-Based Cohort Study in Taiwan," *Journal of Alzheimer's Disease* 44, no. 2 (2015): 573–84, doi:0.3233/JAD-140855.

20 Lilian Calderón-Garcidueñas et al., "Air Pollution and Brain Damage," *Toxicologic Pathology* 30, no. 3 (April 2002): 373–89, https://doi.org/10.1080/01926230252929954.

21 Lilian Calderón-Garcidueñas et al., "Neuroinflammation, Hyperphosphorylated Tau, Diffuse Amyloid Plaques, and Down-Regulation of the Cellular Prion Protein in Air Pollution Exposed Children and Young Adults," *Journal of Alzheimer's Disease* 28, no. 1 (2012): 93–107, doi:10.3233/JAD-2011-110722.

22 Lilian Calderón-Garcidueñas, Maricela Franco-Lira, Antonieta Mora-Tiscareño, Humberto Medina-Cortina, Ricardo Torres-Jardón, and Michael Kavanaugh, "Early Alzheimer's and Parkinson's Disease Pathology in Urban Children: Friend versus Foe Responses—It Is Time to Face the Evidence," *BioMed Research International*, vol. 2013, Article ID 161687, http://dx.doi.org/10.1155/2013/161687.

23 Lilian Calderón-Garcidueñas, Antonieta Mora-Tiscareño, Maricela Franco-Lira, Hongtu Zhu, Zhaohua Lu, Edelmira Solorio, Ricardo Torres-Jardón, and Amadeo D'Angiulli, "Decreases in Short Term Memory, IQ, and Altered Brain Metabolic Ratios in Urban Apolipoprotein ε4 Children Exposed to Air Pollution," *Journal of Alzheimer's Disease* 45, no. 3 (2015): 757–70, doi:10.3233/JAD-142685.

24   Jiu-Chiuan Chen, X. Wang, G. A. Wellenius, M. L. Serre, I. Driscoll, R. Casa-nova, J. J. McArdle, J. E. Manson, H. C. Chui, and M. A. Espeland, "Ambient Air Pollution and Neurotoxicity on Brain Structure: Evidence from Women's Health Initiative Memory Study," *Annals of Neurology* 78, no. 3 (2015): 466–76, doi:10.1002/ana.24460.

25   Jennifer Weuve, "Exposure to Particulate Air Pollution and Cognitive De-cline in Older Women," *Archives of Internal Medicine* 172, no. 3 (2012): 219, doi:10.1001/archinternmed.2011.683.

26   Liuhua Shi, Antonella Zanobetti, Itai Kloog, Brent A. Coull, Petros Koutrakis, Steven J. Melly, and Joel D. Schwartz, "Low-Concentration PM2.5 and Mor-tality: Estimating Acute and Chronic Effects in a Population-Based Study," *Environmental Health Perspectives* 124, no. 1 (2015), doi:10.1289/ehp.1409111.

27   Douglas Dockery, C. Arden Pope III, Xiping Xu, John D. Spengler, James H. Ware, Martha E. Fay, Benjamin G. Ferris, Jr., and Frank E. Speizer, "An Associa-tion between Air Pollution and Mortality in Six U.S. Cities," *New England Jour-nal of Medicine* 329, no. 24 (1993): 1753–59, doi:10.1056/NEJM199312093292401; Douglas Dockery, "Landmark Air Pollution Study Turns 20," interview by Karen Feldscher, January 7, 2014, Harvard T. H. Chan School of Public Health (website), accessed November 9, 2017, https://www.hsph.harvard.edu/news/features/six-cities-air-pollution-study-turns-20/.

28   C. Arden Pope III, Majid Ezzati, and Douglas W. Dockery, "Fine-Particulate Air Pollution and Life Expectancy in the United States," *New England Journal of Medicine* 360, no. 4 (2009): 376–86, doi:10.1056/NEJMsa0805646.

29   Victor Frutos, Mireia González-Comadrán, Ivan Solà, Benedicte Jacquemin, Ramón Carreras, and Miguel A. Checa Vizcaíno, "Impact of Air Pollution on Fertility: A Systematic Review," *Gynecological Endocrinology* 31, no. 1 (2015): 7–13, http://dx.doi.org/10.3109/09513590.2014.958992.

30   Julia E. Heck, Jun Wu, Travis J. Meyers, Michelle Wilhelm, Myles Cockburn, and Beate Ritz, "Abstract 2531: Childhood Cancer and Traffic-Related Air Pol-lution Exposure in Pregnancy and Early Life," *Cancer Research* 73, no. 8 (2013): 2531; Anshu Shrestha, Beate Ritz, Michelle Wilhelm, Jiaheng Qiu, Myles Cock-burn, and Julia E. Heck, "Prenatal Exposure to Air Toxics and Risk of Wilms' Tumor in 0- to 5-Year Old Children," *Journal of Occupational and Environmen-tal Medicine* 56, no. 6 (June 2014): 573–78, doi:10.1097/JOM.0000000000000167.

31   Beate Ritz, Michelle Wilhelm, and Yingxu Zhao, "Air Pollution and Infant Death in Southern California, 1989–2000," *Pediatrics* 118, no. 2 (2006): 493–502, doi:10.1542/peds.2006-0027.

32   Beate Ritz, Fei Yu, Scott Fruin, Guadalupe Chapa, Gary M. Shaw, and John A. Harris, "Ambient Air Pollution and Risk of Birth Defects in Southern Cali-fornia," *American Journal of Epidemiology* 155, no. 1 (2002): 17–25, doi:10.1093/aje/155.1.17: Warren Robak, "Urban Air Pollution Linked to Birth De-fects for First Time; UCLA Research Links Two Pollutants to Increased Risk of Heart Defects," UCLA Newsroom, December 31, 2001, accessed November 9, 2017, http://newsroom.ucla.edu/releases/Urban-Air-Pollution-Linked-to-Birth-2932.

33   Tracy Ann Becera, Michelle Wilhelm, Jørn Olsen, Myles Cockburn, and Beate Ritz, "Ambient Air Pollution and Autism in Los Angeles County, California,"

Environmental Health Perspectives 121, no. 3 (March 2013): 380–86, http:// dx.doi.org/10.1289/ehp.1205827.

34   W. James Gauderman, Robert Urman, Edward Avol, Kiros Berhane, Rob McConnell, Edward Rappaport, Roger Chang, Fred Lurmann, and Frank Gilliland, "Association of Improved Air Quality with Lung Development in Children," *New England Journal of Medicine* 372, no. 10 (2015): 905–13, doi:10.1056 /NEJMoa1414123.

35   "Are Busy Roads Dangerous?" Environmental Research Group, King's College London, London Air Quality Network—King's College London Guide, accessed November 9, 2017, http://www.londonair.org.uk/londonair/guide /BusyRoad.aspx.

Chapter Two

1   Yuanyuan Shi et al., "Nanoscale Characterization of PM2.5 Airborne Pollutants Reveals High Adhesiveness and Aggregation Capability of Soot Particles," *Scientific Reports* 5, no. 1 (2015), doi:10.1038/srep11232; Gladis Labrada-Delgado, Antonio Aragon-Pina, Arturo Campos-Ramos, Telma Castro-Romero, Omar Amador-Munoz, and Rafael Villalobos-Pietrini, "Chemical and Morphological Characterization of PM2.5 Collected during MILAGRO Campaign Using Scanning Electron Microscopy," *Atmospheric Pollution Research* 3, no. 3 (2012): 289–300, doi:10.5094/apr.2012.032.

2   "Explore the Data," State of Global Air 2018.

3   WHO Global Ambient Air Quality Database (update 2018), World Health Organization, accessed August 20, 2018, http://www.who.int/airpollution/data/cities/en/.

4   "What the Study Found: 2 Times More Asthma, 3 Times More Severe Lung Disorder," *Indian Express*, April 2, 2015, https://indianexpress.com/article/india /india-others/what-the-study-found-2-times-more-asthma-3-times-more -severe-lung-disorder/.

5   "Country Profiles: India," Institute for Health Metrics and Evaluation, 2018, accessed June 12, 2018, http://www.healthdata.org/india.

6   MyLinh Duong et al., "Global Differences in Lung Function by Region (PURE): An International, Community-Based Prospective Study," *Lancet Respiratory Medicine* 1, no. 8 (October 2013): 599–609, http://dx.doi.org/10.1016/S2213 -2600(13)70164-4.

7   Gardiner Harris, "Beijing's Bad Air Would Be Step Up for Smoggy Delhi," *New York Times*, January 25, 2014, https://www.nytimes.com/2014/01/26/world/asia /beijings-air-would-be-step-up-for-smoggy-delhi.html.

8   Dr. Sundeep Salvi, director of Chest Research Foundation of Pune, Skype interview with author, April 6, 2015.

9   The World Bank and the Institute for Health Metrics and Evaluation, *The Cost of Air Pollution*.

10   Beth Gardiner, "'My Children Are Suffering but What Can I Do?' Delhi's Polluted Air, by the People Who Live There," Guardian Cities, *Guardian*, June 25, 2015, https://www.theguardian.com/cities/2015/jun/25/delhi-india-pollution -air-quality-most-polluted-city-children-suffering.

11   Press Trust of India, "Arvind Kerjival Leaves for Bangalore to Undergo Naturopathy Treatment," *Indian Express*, March 5, 2015, http://indianexpress.com/article/cities/delhi/arvind-kejriwal-left-for-bangalore-to-undergo-naturopathy-treatment/.

12   Joshua Schulz Apte, "Human Exposure to Urban Vehicle Emissions," PhD diss., University of California, Berkeley, 2013, 1, http://digitalassets.lib.berkeley.edu/etd/ucb/text/Apte_berkeley_0028E_13739.pdf.

13   Gardiner, "'My Children are Suffering.'"

14   *Report on Priority Measures to Reduce Air Pollution and Protect Public Health*, Environment Pollution (Prevention & Control) Authority for the National Capital Region, p. 2, accessed August 20, 2018, https://cdn.cseindia.org/userfiles/EPCA_report.pdf.

15   Vishnu Mathur, director general, Society of Indian Automobile Manufacturers, interview with author, April 9, 2015.

16   Debi Goenka and Sarath Guttikunda, *Coal Kills: An Assessment of Death and Disease Caused by India's Dirtiest Energy Source*, Conservation Action Trust, UrbanEmissions.info and Greenpeace India, March 2013, p. 1, http://www.greenpeace.org/india/Global/india/report/Coal_Kills.pdf.

17   *Air Quality Assessment, Emission Inventory & Source Apportionment Study for Delhi*, National Environmental Engineering Research Institute, December 2008, Table E.2.

18   Pritha Chatterjee, "Vehicle Exhaust, Dust, What Fouls the Air the Most? Studies Disagree," *Indian Express*, January 6, 2016, http://indianexpress.com/article/explained/vehicle-exhaust-dust-what-fouls-the-air-the-most-studies-disagree/; Mukesh Sharma and Onkar Dikshit, *Comprehensive Study on Air Pollution and Green House Gases (GHGs) in Delhi*, Delhi Pollution Control Committee and Department of Environment, NCT Delhi, January 2016, vii, http://delhi.gov.in/DoIT/Environment/PDFs/Final_Report.pdf.

19   Oliver Wainwright, "Blood Bricks: How India's Urban Boom Is Built on Slave Labour," *Guardian*, January 8, 2014, https://www.theguardian.com/artanddesign/architecture-design-blog/2014/jan/08/blood-bricks-india-urbanisation-human-rights-slave-labour; Humphrey Hawksley, "Why India's Brick Kiln Workers 'Live Like Slaves,'" BBC News, January 2, 2014, http://www.bbc.co.uk/news/world-asia-india-25556965.

20   Sarath Guttikunda, director of UrbanEmission.info, phone interview with author, April 8, 2015.

21   Aniruddha Ghosal, Mayura Janwalkar, and Pritha Chatterjee, "While You Are Sleeping: 80,000 Trucks Enter Delhi Every Night, Poison on Wheels," *Indian Express*, April 3, 2015, http://indianexpress.com/article/india/india-others/while-you-are-sleeping-80000-trucks-enter-delhi-every-night-poison-on-wheels/.

22   Gardiner, "'My Children Are Suffering.'"

23   "GDP per Capita (current US$)," World Bank, accessed November 14, 2017, https://data.worldbank.org/indicator/NY.GDP.PCAP.CD.

24   Ramanan Laxminarayan, Public Health Foundation of India, interview with author, April 8, 2015.

25   "CO_2 Emissions (Metric Tons per Capita)," World Bank, accessed November 14, 2017, https://data.worldbank.org/indicator/EN.ATM.CO2E.PC.

26   Gardiner Harris, "Holding Your Breath in India," *New York Times*, May 29, 2015, https://www.nytimes.com/2015/05/31/opinion/sunday/holding-your-breath -in-india.html?_r=0.

27   Gardiner, "'My Children Are Suffering.'"

Chapter Three

1   European Environment Agency, *Spatial Assessment of PM10 and Ozone Concentrations in Europe (2005)*, March 24, 2009, 20, https://www.eea.europa .eu/publications/spatial-assessment-of-pm10-and-ozone-concentrations-in -europe-2005-1.

2   "What Is the Real Health Impact of Poor Air Quality in Greater London?" Clean Air in London, April 19, 2009, accessed November 15, 2017, https:// cleanair.london/health/what-is-the-real-health-impact-of-poor-air-quality -in-greater-london/.

3   "Memorandum Submitted by the Campaign for Clean Air in London (CCAL) (AQ 18)," House of Commons Environmental Audit Committee, December 13, 2009, accessed November 15, 2017, https://publications.parliament.uk/pa /cm200910/cmselect/cmenvaud/229/229we14.htm.

4   "Examination of Witnesses (Question Numbers 57-79)," House of Commons Environmental Audit Committee, February 9, 2010, accessed November 15, 2017, https://publications.parliament.uk/pa/cm200910/cmselect/cmenvaud /229/10020907.htm.

5   Brian G. Miller, *Report on Estimation of Mortality Impacts of Particulate Air Pollution in London*, Institute of Occupational Medicine, June 2010, http:// www.aef.org.uk/uploads/IomReport_1.pdf.

6   "Deaths Registered by Area," Office for National Statistics.

7   C. Arden Pope III, Richard T. Burnett, and Michael J. Thun, "Lung Cancer, Cardiopulmonary Mortality, and Long-Term Exposure to Fine Particulate Air Pollution," *Journal of the American Medical Association* 287, no. 9 (2002): 1132, doi:10.1001/jama.287.9.1132.

8   Committee on the Medical Effects of Air Pollutants, *The Mortality Effects of Long-Term Exposure to Particulate Air Pollution in the United Kingdom*, 2010, https://www.gov.uk/government/uploads/system/uploads/attachment_data /file/304641/COMEAP_mortality_effects_of_long_term_exposure.pdf.

9   Royal College of Physicians, Royal College of Paediatrics and Child Health, *Every Breath We Take: The Lifelong Impact of Air Pollution*, February 2016, https://www.rcplondon.ac.uk/projects/outputs/every-breath-we-take-lifelong -impact-air-pollution.

10   "Explore the Data," State of Global Air 2018.

11   European Environment Agency, *Air Quality in Europe—2017 Report*.

12   "Explore the Data," State of Global Air 2018.

13   "Five Facts About . . . Cars," Office for National Statistics, September 22, 2016, accessed November 15, 2017, http://visual.ons.gov.uk/five-facts-about-cars/.

14   Christine L. Corton, *London Fog: The Biography* (Cambridge, MA: The Belknap Press of Harvard University Press, 2015).

15   *World Health Statistics 2017*, World Health Organization, 66, http://apps.who
      .int/iris/bitstream/10665/255336/1/9789241565486-eng.pdf?ua=1.

16   Steven R. H. Barrett, Rex E. Britter, and Ian A. Waitz, "Global Mortality Attrib-
      utable to Aircraft Cruise Emissions," *Environmental Science & Technology* 44,
      no. 19 (September 1, 2010), doi:10.1021/es101325r.

17   Brady Dennis, "Here's How Much of the Arctic You're Personally Responsible
      for Melting," *Washington Post*, November 3, 2016, https://www.washington
      post.com/news/energy-environment/wp/2016/11/03/heres-how-much-of-the
      -arctic-youre-personally-responsible-for-melting/?utm_term=.04ea1a8d93ad.

18   International Council on Clean Transportation, "Declining Diesel Car Share
      Not a Hurdle for Meeting the European Union's CO_2 Reduction Targets," news
      release, May 7, 2017, accessed November 15, 2017, http://www.theicct.org/news
      /press-release-declining-diesel-car-share-EU.

19   Society of Motor Manufacturers and Traders, *New Car CO2 Report 2014*, 16,
      accessed August 20, 2018, https://www.smmt.co.uk/wp-content/uploads
      /sites/2/SMMT-New-Car-CO2-Report-2014-final1.pdf.

20   "Share of Diesel in New Passenger Cars," ACEA European Automobile Manu-
      facturers Association, accessed November 15, 2017, http://www.acea.be/statistics
      /tag/category/share-of-diesel-in-new-passenger-cars.

21   "Composition of Diesel and Non-Diesel Fleet: 2014," United States Department
      of Transportation, Bureau of Transportation Statistics, October 2015, accessed
      November 15, 2017, https://www.rita.dot.gov/bts/sites/rita.dot.gov.bts/files
      /publications/bts_fact_sheets/oct_2015/html/figure_01.html.

22   International Council on Clean Transportation, "Declining Diesel Car
      Share"; Society of Motor Manufacturers and Traders, *New Car CO2 Report
      2014*.

23   Michel Cames and Eckard Helmers, "Critical Evaluation of the European
      Diesel Car Boom—Global Comparison, Environmental Effects and Various
      National Strategies," *Environmental Sciences Europe* 25, no. 15 (June 22, 2013),
      doi:10.1186/2190-4715-25-15.

24   Sadiq Khan, "Sadiq Khan: Today Is the Day We Really Start to Clean Up Our
      Toxic Air," *Evening Standard*, October 23, 2017, https://www.standard.co.uk
      /comment/comment/today-is-the-day-we-really-start-to-clean-up-our-toxic
      -air-a3665541.html.

25   Damian Carrington, "London Reaches Legal Air Pollution Limit Just One
      Month into the New Year," *Guardian*, January 30, 2018, https://www.theguard
      ian.com/uk-news/2018/jan/30/london-reaches-legal-air-pollution-limit-just
      -one-month-into-the-new-year.

Chapter Four

1   WHO Global Ambient Air Quality Database (Update 2018), World Health
     Organization.

2   Beth Gardiner, "'The Air Is Stinking, It's Dirty': The Fight against Pollution in
     Krakow," *Guardian*, April 13, 2015, https://www.theguardian.com/cities/2015
     /apr/13/air-dirty-fight-pollution-krakow-poland-ban-wood-coal.

3   Beth Gardiner, "Coal in Poland Lowering Life Spans," *New York Times*, June 7, 2015, https://www.nytimes.com/2015/06/08/business/energy-environment /coal-in-poland-lowering-life-spans.html?_r=1.

4   "Coal," International Energy Agency, accessed November 16, 2017, https://www .iea.org/topics/coal/.

5   "How the Coal Industry Fuels Climate Change," Greenpeace International, July 1, 2016, accessed November 16, 2017, http://www.greenpeace.org/international /en/campaigns/climate-change/coal/Coal-fuels-climate-change/.

6   Gardiner, "'The Air Is Stinking.'"

7   Piotr Ziarkowski, interview with author, March 7, 2015.

8   Dan Fagin, *Toms River: A Story of Science and Salvation* (New York: Bantam Books, 2013), 47–49.

9   Gardiner, "'The Air Is Stinking.'"

10   Gardiner, "'The Air Is Stinking.'"

11   European Environment Agency, *Air Quality in Europe—2017 Report*, p. 57.

12   "Economic Cost of the Health Impact of Air Pollution in Europe," World Health Organization Regional Office for Europe, Organization for Economic Co-operation and Development, 2015, http://www.euro.who.int/__data /assets/pdf_file/0004/276772/Economic-cost-health-impact-air-pollution-en .pdf?ua=1.

13   Julia Michalak, demosEuropa, interview with author, March 11, 2015; Gardiner, "Coal in Poland."

14   WWF European Policy Office, CAN Europe, HEAL, and Sandbag, *Europe's Dark Cloud: How Coal-Burning Countries Are Making Their Neighbours Sick*, Brussels, Belgium, June 2017, p. 17, http://env-health.org/IMG/pdf/dark_cloud -full_report_final.pdf.

15   Gardiner, "Coal in Poland."

16   Michal Olszewski, Krakow bureau chief, *Gazeta Wyborcza*, interview with author, March 4, 2015.

Chapter Five

1   "California's Central Valley," California Water Science Center, U.S. Geological Survey, accessed November 16, 2017, https://ca.water.usgs.gov/projects/central -valley/about-central-valley.html.

2   American Lung Association, *State of the Air 2018*.

3   "Welcome to Central Valley Almond Association (CVAA)," Central Valley Almond Association, accessed November 16, 2017, http://www.centralvalleyal mond.com.

4   Michael Pollan, "Our Decrepit Food Factories," *New York Times*, December 12, 2007, http://www.nytimes.com/2007/12/16/magazine/16wwln-lede-t.html.

5   Beth Gardiner, "How Growth in Dairy Is Affecting the Environment," *New York Times*, May 1, 2015, https://www.nytimes.com/2015/05/04/business/energy -environment/how-growth-in-dairy-is-affecting-the-environment.html?_r=0.

6   "California Drought: Livestock, Dairy, and Poultry Sectors," U.S. Department of Agriculture Economic Research Service, June 14, 2017, accessed November 16,

2017, http://usda.proworks.com/topics/in-the-news/california-drought-farm-and-food-impacts/california-drought-livestock-dairy-and-poultry-sectors/.

7   "U.S. Methane 'Hot Spot' Bigger than Expected," Science@NASA, October 9, 2014, accessed November 16, 2017, https://science.nasa.gov/science-news/science-at-nasa/2014/09oct_methanehotspot/.

8   "2016 Premature Birth Report Card," March of Dimes, accessed November 16, 2017, https://www.marchofdimes.org/materials/premature-birth-report-card-california.pdf.

9   Gardiner, "How Growth in Dairy Is Affecting the Environment."

10   "Conversation with Mark Arax," C-SPAN.org, January 11, 2017, accessed May 7, 2018, https://www.c-span.org/video/?423503-1/conversation-mark-arax.

11   Mark Arax, "A Kingdom from Dust," *California Sunday Magazine*, January 31, 2018, https://story.californiasunday.com/resnick-a-kingdom-from-dust; Mark Bittman, "Everyone Eats There," *New York Times Magazine*, October 10, 2012, https://www.nytimes.com/2012/10/14/magazine/californias-central-valley-land-of-a-billion-vegetables.html.

12   "NASA Data Show California's San Joaquin Valley Still Sinking," NASA, February 28, 2017, accessed May 7, 2018, https://www.nasa.gov/feature/jpl/nasa-data-show-californias-san-joaquin-valley-still-sinking.

13   Susanne E. Bauer, Kostas Tsigaridis, and Ron Miller, "Significant Atmospheric Aerosol Pollution Caused by World Food Cultivation," *Geophysical Research Letters* 43, no. 10 (May 16, 2016), doi:10.1002/2016gl068354.

14   J. Lelieveld, J. S. Evans, M. Fnais, D. Giannadaki, and A. Pozzer, "The Contribution of Outdoor Air Pollution Sources to Premature Mortality on a Global Scale," *Nature* 525, no. 09 (September 17, 2015): 367–71, doi:10.1038/nature15371.

15   M. Vieno et al., "The UK Particulate Matter Air Pollution Episode of March–April 2014: More Than Saharan Dust," *Environmental Research Letters* 11, no. 4 (March 24, 2016), doi:10.1088/1748-9326/11/4/044004.

16   "Fracking Threatens Health of Kern County Communities Already Overburdened with Pollution," Natural Resources Defense Council, September 2014, accessed November 16, 2017, https://www.nrdc.org/sites/default/files/california-fracking-risks-kern-FS.pdf.

17   "Conversation with Mark Arax," C-SPAN.org.

18   "CalEnviroScreen 3.0 Data," California Office of Environmental Health Hazard Assessment, January 30, 2017, accessed November 16, 2017, https://oehha.ca.gov/calenviroscreen/maps-data/download-data.

19   California Environmental Protection Agency Air Resources Board, *Report on Arvin Special Purpose PM2.5 Monitoring Project*, June 2011, https://www.arb.ca.gov/aaqm/Special%20Monitoring%20in%20Arvin3.pdf.

20   Adam Herbets, "Arvin Community Services District Penalized for Arsenic Levels in Water," Bakersfieldnow.com, KBAK/KBFX, October 6, 2015, accessed November 16, 2017, http://bakersfieldnow.com/news/health/arvin-community-services-district-penalized-for-arsenic-levels-in-water.

21   Jane V. Hall and Victor Brajer, *The Benefits of Meeting Federal Clean Air Standards in the South Coast and San Joaquin Valley Air Basins*, November 2008, http://publichealth.lacounty.gov/mch/asthmacoalition/docs/BenefitsofMeetingCleanAirStandards_11_06_08.pdf.

22   David Lighthall, San Joaquin Valley Air Pollution Control District, interview with author, February 4, 2015.

Chapter Six

1   "Household Air Pollution and Health," World Health Organization, accessed November 9, 2017, http://www.who.int/mediacentre/factsheets/fs292/en/.
2   World Health Organization, "9 out of 10 People" and "Household Air Pollution and Health."
3   Donate to the Masekos at https://stekakids.com.
4   Caroline A. Ochieng, Cathryn Tonne, Sotiris Vardoulakis, and Jan Semenza, "Household Air Pollution in Low and Middle Income Countries," *Oxford Research Encyclopedia of Environmental Science*, December 2017, doi:10.1093 /acrefore/9780199389414.013.25.
5   Kevin Mortimer et al., "A Cleaner Burning Biomass-Fuelled Cookstove Intervention to Prevent Pneumonia in Children under 5 Years Old in Rural Malawi (the Cooking and Pneumonia Study)," *Lancet* 389, no. 10065 (January 14, 2017): 167–75, doi:10.1016/s0140-6736(16)32507-7.
6   Ambuj Sagar, Kalpana Balakrishnan, Sarath Guttikunda, Anumita Roychowdhury, and Kirk R. Smith, "India Leads the Way: A Health-Centered Strategy for Air Pollution," *Environmental Health Perspectives* 124, no. 7 (2016), doi:10 .1289/ehp90.
7   Beth Gardiner, "The Challenges of Cleaning Up Cooking," *New York Times*, December 8, 2015, https://www.nytimes.com/2015/12/09/business/energy -environment/the-challenges-of-cleaning-up-cooking.html.
8   Sagar et al., "India Leads the Way."
9   Dr. Narendra Arora, INCLEN Trust International, interview with author, April 9, 2015.
10   Kirk R. Smith, J. P. McCracken, M. W. Weber, A. Hubbard, A. Jenny, L. M. Thompson, J. Balmes, A. Diaz, B. Arana, and N. Bruce, "Effect of Reduction in Household Air Pollution on Childhood Pneumonia in Guatemala (RESPIRE): A Randomised Controlled Trial," *Lancet* 378, no. 9804 (November 12, 2011), doi:10.1016/s0140-6736(11)60921-5.
11   Smith told me this story during a phone conversation following my visit to his office; I have incorporated it and other material from that follow-up interview into my account of our in-person meeting for the sake of narrative smoothness.
12   Gardiner, "The Challenges of Cleaning Up Cooking."
13   Ochieng et al., "Household Air Pollution."
14   Suani T. Coelho and José Goldemberg, "Energy Access: Lessons Learned in Brazil and Perspectives for Replication in Other Developing Countries," *Energy Policy* 61 (June 20, 2013), doi:10.1016/j.enpol.2013.05.062.
15   Ochieng et al., "Household Air Pollution."
16   Professor Kalpana Balakrishnan, Department of Environmental Health Engineering, Sri Ramachandra University, Skype interview with author, April 10, 2015.

17   Dorothy L. Robinson, "Response: 2.4 Times More PM2.5 Pollution from Domestic Wood Burning than Traffic," *BMJ*, January 27, 2016, https://www.bmj.com/content/350/bmj.h2757/rr-1.

18   World Health Organization Regional Office for Europe, *Residential Heating with Wood and Coal: Health Impacts and Policy Options in Europe and North America*, Copenhagen, Denmark, 2015, http://www.euro.who.int/__data/assets/pdf_file/0009/271836/ResidentialHeatingWoodCoalHealthImpacts.pdf?ua=1.

19   Gary Fuller, "Pollutionwatch: Wood Burning Worsening UK Air Quality," *Guardian*, February 1, 2018, https://www.theguardian.com/environment/2018/feb/01/pollutionwatch-wood-burning-worsening-uk-air-quality.

20   Leigh Crilley, University of Birmingham, phone interview with author, January 31, 2018. Quoted in Beth Gardiner, "Is Your Wood Stove Choking You? How Indoor Fires Are Suffocating Cities," *Guardian*, February 22, 2018, https://www.theguardian.com/cities/2018/feb/22/wood-diesel-indoor-stoves-cities-pollution.

21   John D. Sterman, Lori Siegel, and Juliette N. Rooney-Varga, "Does Replacing Coal with Wood Lower CO_2 Emissions? Dynamic Lifecycle Analysis of Wood Bioenergy," *Environmental Research Letters* 13, no. 1 (January 18, 2018), doi:10.1088/1748-9326/aaa512.

Chapter Seven

1   Leon Billings, transcript of interview with Don Nicoll, Edmund S. Muskie Oral History Collection, Bates College Digital Library, November 29, 2001, http://digilib.bates.edu/collect/muskieor/index/assoc/HASH9799.dir/doc.pdf.

2   John Freshman, senior director of governmental affairs, Best Best & Krieger, interview with author, April 7, 2017.

3   Jim Dwyer, "Remembering a City Where the Smog Could Kill," *New York Times*, February 28, 2017, https://www.nytimes.com/2017/02/28/nyregion/new-york-city-smog.html?_r=0; Daniel C. Walsh, Steven N. Chillrud, H. James Simpson, and Richard F. Bopp, "Refuse Incinerator Particulate Emissions and Combustion Residues for New York City during the 20th Century," *Environmental Science & Technology* 35, no. 12 (June 15, 2001), doi:10.1021/es0013475.

4   "500% Rise in Emphysema Mortality Rate in Last Decade Reported for City," *New York Times*, March 12, 1970, https://www.nytimes.com/1970/03/12/archives/500-rise-in-emphysema-mortality-rate-in-last-decade-reported-for.html.

5   "Edmund S. Muskie, Late a Senator from Maine, Memorial Tributes in the Congress of the United States," S. Doc. 104–17 (1996), https://www.gpo.gov/fdsys/pkg/CDOC-104sdoc17/pdf/CDOC-104sdoc17.pdf.

6   Freshman, interview with author.

7   John C. Whitaker, "Earth Day Recollections: What It Was Like When the Movement Took Off," *EPA Journal*, U.S. Environmental Protection Agency, August 1988, accessed November 17, 2017, https://archive.epa.gov/epa/aboutepa/earth-day-recollections-what-it-was-when-movement-took.html.

8   Maxmillian Angerholzer III, James Kitfield, Norman Ornstein, and Stephen Skowronek, eds., *Triumphs and Tragedies of the Modern Presidency: Case Studies in Presidential Leadership*, 2nd ed. (Santa Barbara: Praeger, 2016), 191.

9   *The Benefits and Costs of the Clean Air Act, 1970 to 1990*, prepared for U.S. Congress by U.S. Environmental Protection Agency, October 1997, https://www .epa.gov/sites/production/files/2015-06/documents/contsetc.pdf.

10   David Corn, "Why Henry Waxman Was One of the Most Important Congressmen Ever," *Mother Jones*, January 30, 2014, http://www.motherjones.com/mojo /2014/01/henry-waxman-retires-accomplishments.

11   Jonathan Cohn, "Farewell to Henry Waxman, a Liberal Hero," *New Republic*, January 31, 2014, https://newrepublic.com/article/116418/henry-waxman -retiring-heres-why-well-miss-him.

12   Henry A. Waxman with Joshua Green, *The Waxman Report: How Congress Really Works* (New York: Twelve, 2010), 2.

13   Nancy E. Roman, "GOP Feels Bite of 'Bulldog' Waxman," *Washington Times*, May 6, 1998, viewed at https://kaiserhealthnews.files.wordpress.com/2014/01 /news_profiles_bite_bulldog_5_6_98.pdf.

14   Roman, "GOP Feels Bite."

15   Harold Meyerson, "Henry Waxman, Liberalism's Legislative Genius," *Washington Post*, February 5, 2014, https://www.washingtonpost.com/opinions /harold-meyerson-henry-waxman-liberalisms-legislative-genius/2014/02/05 /b84daa00-8dd7-11e3-98ab-fe5228217bd1_story.html?utm_term=.208e8f7bf314.

16   Waxman with Green, *Waxman Report*, 78.

17   "Tactics of an Ace in the Congressional Air Wars," *Washington Post*, December 14, 1982, https://www.washingtonpost.com/archive/politics/1982/12/14 /tactics-of-an-ace-in-the-congressional-air-wars/d05d2f82-22b3-4db7-93b7 -c944c6466bb1/?utm_term=.cc2ddf16b540.

18   "Required Reading: Character Study," *New York Times*, October 21, 1981, http:// www.nytimes.com/1981/10/21/us/required-reading-character-study.html.

19   Ruth Marcus, "21–20 Vote in Committee Preserves Clean Air Act from Industry Changes," *Washington Post*, August 12, 1982, https://www.washingtonpost .com/archive/politics/1982/08/12/21-20-vote-in-committee-preserves-clean -air-act-from-industry-changes/787b4fdd-6d5f-42b1-99fb-41f9e791e9b6/?utm _term=.78f57e0d7668; Waxman with Green, Waxman Report, 82–85.

20   William Robbins, "House Panel Not Reassured by Carbide Officials," *New York Times*, December 15, 1984, http://www.nytimes.com/1984/12/15/world/house -panel-not-reassured-by-carbide-officials.html.

21   George Lobsenz, "Art of Compromise Breathes Life into Clean Air Bill," United Press International, April 7, 1990, https://www.upi.com/Archives/1990/04/07 /Art-of-compromise-breathes-life-into-clean-air-bill/4418639460800/; Waxman with Green, *Waxman Report*, 99–100.

22   Jean Chemnick and Manuel Quiñones, "Air Pollution: Clean Air Act Debates Show How Much Politics Have Changed," *E&E News*, February 13, 2014, https://www.eenews.net/stories/1059994544.

23   *Benefits and Costs of the Clean Air Act 1990–2020, the Second Prospective Study*, April 2011, Environmental Protection Agency, accessed November 17, 2017, https://www.epa.gov/clean-air-act-overview/benefits-and-costs-clean-air-act -1990-2020-second-prospective-study.

24   Christine Todd Whitman, *It's My Party Too: The Battle for the Heart of the GOP and the Future of America* (New York: Penguin Press, 2005), 170–71.

25   Whitman, *It's My Party Too*, 175–76; Douglas Jehl with Andrew C. Revkin, "Bush, in Reversal, Won't Seek Cut In Emissions of Carbon Dioxide," *New York Times*, March 14, 2001, http://www.nytimes.com/2001/03/14/us/bush-in-reversal-won-t-seek-cut-in-emissions-of-carbon-dioxide.html.

26   George W. Bush, letter to members of the Senate on the Kyoto Protocol on climate change, American Presidency Project, March 13, 2001, accessed November 17, 2017, http://www.presidency.ucsb.edu/ws/?pid=45811.

27   Linda Greenhouse, "Justices Say E.P.A. Has Power to Act on Harmful Gases," *New York Times*, April 3, 2007, http://www.nytimes.com/2007/04/03/washington/03scotus.html.

28   "Rep. Pete Olson—Texas District 22," Center for Responsive Politics, accessed November 17, 2017, https://www.opensecrets.org/members-of-congress/industries?cycle=2016&cid=N00029285&type=C&newmem=N.

29   Emily Atkin, "Scott Pruitt Declares War on Air Pollution Science," *New Republic*, October 31, 2017, https://newrepublic.com/article/145582/scott-pruitt-declares-war-air-pollution-science.

30   U.S. Environmental Protection Agency, "Comparison of Growth Areas and Emissions, 1970–2016," accessed August 20, 2018, https://www.epa.gov/sites/production/files/2018-07/2017_baby_graphic_1970-2017.png.

Chapter Eight

1   John Mooney, presentation to California Air Resources Board, October 9, 2007, accessed November 17, 2017, https://www.arb.ca.gov/research/seminars/mooney/mooney.htm.

2   Tom McCarthy, *Auto Mania: Cars, Consumers and the Environment* (New Haven, CT: Yale University Press, 2007), 126.

3   McCarthy, *Auto Mania*, 49–50.

4   Jamie Lincoln Kitman, "The Secret History of Lead," *Nation*, March 2, 2000, https://www.thenation.com/article/secret-history-lead/.

5   McCarthy, *Auto Mania*, 49–51.

6   Philip J. Landrigan, "The Worldwide Problem of Lead in Petrol," *Bulletin of the World Health Organization* 80, no. 10 (2002): 768, http://www.who.int/bulletin/archives/80(10)768.pdf.

7   Lee Iacocca, as quoted in Jack Doyle, *Taken for a Ride: Detroit's Big Three and the Politics of Pollution* (New York: Four Walls Eight Windows, 2000), 69.

8   McCarthy, *Auto Mania*, 177.

9   Joe Colucci, phone interview with author, April 24, 2017.

10   Historical data from National Health and Nutrition Examination Survey, as summarized in "Blood Lead Levels in Children Aged 1–5 Years—United States, 1999–2010," *Morbidity and Mortality Weekly Report*, 62, no. 13 (April 5, 2013): 245–48, https://www.cdc.gov/mmwr/preview/mmwrhtml/mm6213a3.htm; assessment of IQ impact in V. M. Thomas, "The Elimination of Lead in Gasoline," *Annual Review of Energy and the Environment* 20 (1995): 305, doi: https://doi.org/10.1146/annurev.eg.20.110195.001505.

11   Recollections of then-EPA deputy administrator John Quarles, as quoted in Doyle, *Taken for a Ride*, 83–84.

12   *New York Times* ad, March 1972, as reproduced in Doyle, *Taken for a Ride*, 92.

13   Doyle, *Taken For a Ride*, 106.

14   John Mooney, phone interview with author, June 6, 2017.

15   United Nations, "Phase-Out of Leaded Petrol Brings Huge Health and Cost Benefits—UN-backed Study," news release, October 27, 2011, UN News Centre, accessed November 27, 2017, http://www.un.org/apps/news/story .asp?newsID=40226#,WhwhILacau5.

16   Jack Ewing, "Inside VW's Campaign of Trickery," *New York Times*, May 6, 2017, https://www.nytimes.com/2017/05/06/business/inside-vws-campaign-of -trickery.html?_r=0.

17   Ewing, "Inside VW's Campaign of Trickery"; Bill Vlasic and Aaron M. Kessler, "It Took E.P.A. Pressure to Get VW to Admit Fault," *New York Times*, September 21, 2015, https://www.nytimes.com/2015/09/22/business/it-took -epa-pressure-to-get-vw-to-admit-fault.html.

18   U.S. Department of Justice, Office of Public Affairs, "Volkswagen AG Agrees to Plead Guilty and Pay $4.3 Billion in Criminal and Civil Penalties; Six Volkswagen Executives and Employees Are Indicted in Connection with Conspiracy to Cheat U.S. Emissions Tests," news release, January 11, 2017, U.S. Department of Justice, accessed November 27, 2017, https://www.justice.gov/opa/pr /volkswagen-ag-agrees-plead-guilty-and-pay-43-billion-criminal-and-civil -penalties-six; Ewing, "Inside VW's Campaign of Trickery."

19   Peter Mock, "First Look: Results of the German Transport Ministry's Post-VW Vehicle Testing," International Council on Clean Transportation, April 25, 2016, accessed November 27, 2017, http://www.theicct.org/blog/staff/first-look -results-german-transport-ministrys-post-vw-vehicle-testing.

20   Damian Carrington, Gwyn Topham, and Peter Walker, "Revealed: Nearly All New Diesel Cars Exceed Official Pollution Limits," *Guardian*, April 23, 2016, https://www.theguardian.com/business/2016/apr/23/diesel-cars-pollution -limits-nox-emissions.

21   "Composition of Diesel and Non-Diesel Fleet: 2014," U.S. Department of Transportation, Bureau of Transportation Statistics, October 2015, accessed November 27, 2017, https://www.rita.dot.gov/bts/sites/rita.dot.gov.bts/files /publications/bts_fact_sheets/oct_2015/html/figure_01.html.

22   Susan C. Anenberg et al., "Impacts and Mitigation of Excess Diesel-Related NOx Emissions in 11 Major Vehicle Markets," *Nature* 545 (May 15, 2017), doi:10.1038/nature22086.

23   "CO_2 Emissions (Metric Tons per Capita)," World Bank, accessed November 27, 2017, https://data.worldbank.org/indicator/EN.ATM.CO2E.PC.

24   Jack Ewing, "As German Election Looms, Politicians Face Voters' Wrath for Ties to Carmakers," *New York Times*, September 13, 2017, https://www.nytimes .com/2017/09/13/business/germany-diesel-election.html?_r=0.

25   Gernot Heller, "Merkel Says Found Out about VW Dieselgate Scandal from Media," Reuters, March 8, 2017, https://www.reuters.com/article/us -volkswagen-emissions-merkel/merkel-says-found-out-about-vw-dieselgate -scandal-from-media-idUSKBN16F21M.

26   "Progress Cleaning the Air and Improving People's Health," Environmental Protection Agency, accessed November 14, 2017, https://www.epa.gov/clean -air-act-overview/progress-cleaning-air-and-improving-peoples-health.

27   "Number of U.S. Aircraft, Vehicles, Vessels, and Other Conveyances," U.S. Department of Transportation, Bureau of Transportation Statistics, accessed November 27, 2017, https://www.rita.dot.gov/bts/sites/rita.dot.gov.bts/files /publications/national_transportation_statistics/html/table_01_11.html.

28   "Cars on England's Roads Increase by Almost 600,000 in a Year," BBC News, January 20, 2016, http://www.bbc.co.uk/news/uk-england-35312562.

29   "PC World Vehicles in Use," International Organization of Motor Vehicle Manufacturers, accessed November 27, 2017, http://www.oica.net/wp-content /uploads//PC_Vehicles-in-use.pdf.

30   *Transport Outlook: Seamless Transport for Greener Growth*, International Transport Forum at the Organization for Economic Co-Operation and Development 2012: 31, accessed August 20, 2018, https://www.oecd.org/greengrowth /greening-transport/Transport%20Outlook%202012.pdf.

31   David Roberts, "China Made Solar Panels Cheap. Now It's Doing the Same for Electric Buses," *Vox*, April 17, 2018, https://www.vox.com/energy -and-environment/2018/4/17/17239368/china-investment-solar-electric-buses -cost.

32   Kara Swisher, "Elon Musk Is the Id of Tech," *New York Times*, August 16, 2018, https://www.nytimes.com/2018/08/16/opinion/elon-musk-crazy-tesla.html.

33   Swisher, "Elon Musk Is the Id of Tech."

Chapter Nine

1   Chip Jacobs and William J. Kelly, *Smogtown: The Lung-Burning History of Pollution in Los Angeles* (New York: Overlook Press, 2008), 13–17, 35, 51–52.

2   Arthur Winer, emeritus professor, Department of Environmental Health Sciences, UCLA, Skype interview with author, January 27, 2015.

3   Mary Nichols, "UCLA Faculty Voice: How Angelenos Beat Back Smog," UCLA Newsroom, October 20, 2015, accessed November 14, 2017, http://newsroom .ucla.edu/stories/ucla-faculty-voice-how-angelenos-beat-back-smog.

4   Connie Koenenn, "Bent on Clearing the Air: Attorney Mary Nichols Hopes 'Amazing L.A.' Campaign Will Spur Residents to Join the Fight to Keep Los Angeles from Choking to Death," *Los Angeles Times*, November 20, 1991, http:// articles.latimes.com/1991-11-20/news/vw-109_1_los-angeles-attorney.

5   Koenenn, "Bent on Clearing the Air."

6   "Cleaning Up Diesel Trucks in California," NRDC, September 2008, accessed November 15, 2017, https://www.nrdc.org/sites/default/files/cadieselfunding .pdf.

7   Chris Megerian, "Mary Nichols Has 'Rock Star' Influence as Top Air Quality Regulator," *Los Angeles Times*, December 27, 2014, http://beta.latimes.com /local/politics/la-me-pol-adv-mary-nichols-20141228-story.html.

8   Lisa P. Jackson, "The 2013 *Time* 100: Mary Nichols," *Time*, April 18, 2013, http:// time100.time.com/2013/04/18/time-100/slide/mary-nichols/.

9   Tony Barboza, "New Attack on California's Dirty Air," *Los Angeles Times*, October 1, 2015, http://beta.latimes.com/local/lanow/la-me-ln-what-new-smog -rules-mean-for-california-air-pollution-20150930-story.html.

10   Roy Rivenburg, "Well, Well: Oil Rigs Return," *Los Angeles Times*, November 28, 2005, http://articles.latimes.com/2005/nov/28/local/me-oil28; Alan Taylor, "The Urban Oil Fields of Los Angeles," *Atlantic*, August 26, 2014, https://www.theatlantic.com/photo/2014/08/the-urban-oil-fields-of-los-angeles/100799/.

11   Chris Kirkham and Andrew Khouri, "L.A., Long Beach Ports Losing to Rivals amid Struggle with Giant Ships," *Los Angeles Times*, June 2, 2015, http://beta.latimes.com/business/la-fi-big-ships-ports-20150602-story.html.

12   State of California Department of Justice, "Attorney General Kamala D. Harris Seeks to Join Suit to Protect Public Health in Mira Loma," news release, September 8, 2011, accessed November 15, 2017, https://oag.ca.gov/news/press-releases/attorney-general-kamala-d-harris-seeks-join-suit-protect-public-health-mira-loma.

13   "Air Pollution from Ships," Transport & Environment, accessed November 15, 2017, https://www.transportenvironment.org/what-we-do/shipping/air-pollution-ships.

14   Melissa Lin Perella, "A Decade of Progress at Southern California Ports," NRDC, October 1, 2012, accessed November 15, 2017, https://www.nrdc.org/experts/melissa-lin-perrella/decade-progress-southern-california-ports.

15   Diane Bailey, Thomas Plenys, Gina M. Solomon, Todd R. Campbell, Gail Ruderman Feuer, Julie Masters, and Bella Tonkonogy, *Harboring Pollution: Strategies to Clean Up U.S. Ports*, NRDC, August 2004, 25, https://www.nrdc.org/sites/default/files/ports2.pdf.

16   Ruben Garcia, president, Advanced Cleanup Technologies Inc., phone interview with author, March 28, 2016.

17   Tony Barboza, "Ships to Slow Down off California to Save Whales and Cut Pollution," *Los Angeles Times*, August 6, 2014, http://beta.latimes.com/science/sciencenow/la-sci-sn-ships-slow-down-whales-pollution-20140805-story.html.

18   Julia Pyper, "EPA Bans Sooty Ship Fuel Off U.S. Coasts," ClimateWire, *Scientific American*, August 2, 2012, https://www.scientificamerican.com/article/epa-bans-sooty-ship-fuel-off-us-coasts/.

19   South Coast Air Quality Management District, *Multiple Air Toxics Exposure Study in the South Coast Air Basin*, May 2015, ES-2, http://www.aqmd.gov/docs/default-source/air-quality/air-toxic-studies/mates-iv/mates-iv-final-draft-report-4-1-15.pdf?sfvrsn=7.

20   Kirkham and Khouri, "L.A., Long Beach Ports."

21   Mobility 21, "Southern California: The Heart of America's Freight Movement System," Riverside County Transportation Commission, accessed August 20, 2018, http://www.rctcdev.info/uploads/media_items/southern-california-the-heart-of-america-s-freight-movement-system.original.pdf.

22   "About the Port: Facts and Figures," Port of Los Angeles, accessed August 20, 2018, https://www.portoflosangeles.org/about/facts.asp; "Facts at a Glance," Port of Long Beach, accessed August 20, 2018, http://www.polb.com/about/facts.asp.

23   Victor Brajer, Jane V. Hall, and Frederick W. Lurmann, "Valuing Health Effects: The Case of Ozone And Fine Particles in Southern California,"

Contemporary Economic Policy 29, no. 4 (December 15, 2010), doi:10.1111/j.1465
-7287.2010.00240.x.

24   Tony Barboza, "The Port of L.A. Rolled Back Measures to Cut Pollution—
during its 'Green' Expansion," *Los Angeles Times*, December 14, 2014, http://
beta.latimes.com/local/california/la-me-port-pollution-20151215-story.html.

25   Frances Tornabene De Sousa, interview with author, August 26, 2016.

Chapter Ten

1   Emile Dirks, review of *Insight*, by Chai Jing, China Open Research Network
at the University of Toronto, November 26, 2014, http://corn.groups.politics
.utoronto.ca/?p=238.

2   Mengyu Dong, trans., "Translation: Interview With Chai Jing," *China Digital Times*, http://chinadigitaltimes.net/2015/03/translation-peoples-daily
-interview-chai-jing/.

3   "2016 Prison Census: 259 Journalists Jailed Worldwide," Committee to Protect Journalists, accessed November 27, 2017, https://cpj.org/imprisoned
/2016.php.

4   Benjamin Ismail, head of Asia-Pacific desk, Reporters Without Borders, phone
interview with author, March 24, 2017.

5   WHO Global Ambient Air Quality Database (update 2018), World Health
Organization.

6   "Explore the Data," State of Global Air 2018.

7   Joanna Chiu (@joannachiu), "Not Sure What I Think about Toxic
Snow," Twitter, February 21, 2017, https://twitter.com/joannachiu/status
/833967743904804864.

8   Alex Linder, "Beijing Warns Residents Not to Play in 'Very Dirty Snow,'"
Shanghaiist, January 6, 2017, http://shanghaiist.com/2017/01/06/beijing_smog
_snow.php.

9   Beth Gardiner, "Field Notes: Pollution in China," Pulitzer Center on Crisis
Reporting, May 26, 2017, https://pulitzercenter.org/reporting/pollution-china.

10   Steven Q. Andrews, "Beijing's Hazardous Blue Sky," *China Dialogue*, December 5, 2011, https://www.chinadialogue.net/article/show/single/en/4661-Beijing
-s-hazardous-blue-sky.

11   Xinhua News Agency, "Beijing Reaches Annual 'Blue Sky Days' Target," China.
org.cn, December 18, 2011, http://www.china.org.cn/environment/2011-12/18
/content_24184374.htm.

12   "Recent Statistical Revisions Suggest Higher Historical Coal Consumption in
China," U.S. Energy Information Administration, September 16, 2015, accessed
November 27, 2017, https://www.eia.gov/todayinenergy/detail.php?id=22952.

13   Beth Gardiner, "China's Surprising Solutions to Clear Killer Air," *National
Geographic*, May 5, 2017, https://news.nationalgeographic.com/2017/05/china
-air-pollution-solutions-environment-tangshan/.

14   Gardiner, "Field Notes."

15   Christopher Cairns and Elizabeth Plantan, "Why Autocrats Sometimes Relax
Online Censorship of Sensitive Issues: A Case Study of Microblog Discus-

sion of Air Pollution in China," October 9, 2016, http://www.chrismcairns.com /uploads/3/0/2/2/30226899/cairns_and_plantan_—_why_autocrats_some times_relax_online_censorship_of_sensitive_issues.pdf.

16 Health Effects Institute, *State of Global Air 2017*, Boston, Massachusetts, 2017, p. 7, https://www.stateofglobalair.org/sites/default/files/SOGA2017_report.pdf.

17 Reuters staff, "Companies Told to Halt Production in Northern Chinese City in Anti-pollution Drive," Reuters, November 21, 2016, http://uk.reuters.com /article/us-china-industry-pollution-idUKKBN13G1OS.

18 "China's Environmental Ministry Announces the Most and Least Polluted Cities of 2016," *Shanghaiist*, January 23, 2017, http://shanghaiist.com/2017/01/23 /top_smog_cities.php.

19 Gardiner, "China's Surprising Solutions."

20 Joe Sandler Clarke, "China's 'War on Pollution' Leads to Big Improvements in Air Quality," *Unearthed*, Greenpeace UK, July 22, 2015, https://unearthed.green peace.org/2015/07/22/chinas-war-on-pollution-leads-to-big-improvements -in-air-quality/.

21 Michael Greenstone and Patrick Schwarz, *Air Quality Life Index Update March 2018: Is China Winning Its War on Pollution?* Energy Policy Institute at the University of Chicago, https://epic.uchicago.edu/sites/default/files/UCH-EPIC -AQLI_Update_8pager_v04_Singles_Hi%20%282%29.pdf.

22 Lauri Myllyvirta, "China Kept On Smashing Renewables Records in 2016," *Unearthed*, Greenpeace UK, January 6, 2017, https://unearthed.greenpeace. org/2017/01/06/china-five-year-plan-energy-solar-record-2016.

23 Beth Gardiner, "Three Reasons to Believe in China's Renewable Energy Boom," *National Geographic*, May 12, 2017, https://news.nationalgeographic.com/2017/05 /china-renewables-energy-climate-change-pollution-environment/.

Chapter Eleven

1 Gunnar Nehrke, Bundesverband CarSharing, email message to author, July 3, 2017.

2 "Mobility in the City—Berlin Traffic in Figures," Berlin Senate Department for the Environment, Transport and Climate Protection, accessed November 28, 2017, https://www.berlin.de/senuvk/verkehr/politik_planung/zahlen_fakten /entwicklung/index_en.shtml.

3 "Statistical Portrait Frankfurt am Main 2013," City of Frankfurt am Main, accessed November 28, 2017, https://www.frankfurt.de/sixcms/media.php /678/J2014K00_Statistisches_Portrait_2013.pdf.

4 "Cologne Facts & Figures," City of Cologne, October 2016, accessed November 28, 2017, http://www.stadt-koeln.de/mediaasset/content/pdf15/statistik -standardinformationen/cologne_facts_and_figures_2016.pdf.

5 Julie Beck, "The Decline of the Driver's License," *Atlantic*, January 22, 2016, https://www.theatlantic.com/technology/archive/2016/01/the-decline-of-the -drivers-license/425169/; Ann Berrington and Julia Mikolai, *Young Adults' License-Holding and Driving Behaviour in the UK*, Royal Automobile Club Foundation for Motoring, December 2014, accessed November 28, 2017,

http://www.racfoundation.org/assets/rac_foundation/content/downloadables
/Young-Adults-Licence-Holding-Berrington-Mikolai-DEC-2014.pdf.

Epilogue

1   Howard H. Baker, Jr., "Cleaning America's Air—Progress and Challenges,"
     remarks delivered at the University of Tennessee, Knoxville; The Edmund S.
     Muskie Foundation, March 9, 2005, accessed November 29, 2017, http://www
     .muskiefoundation.org/baker.030905.html.

INDEX